Introduction to plant
population ecology

Introduction to plant population ecology

Jonathan W. Silvertown
Lecturer in Biology, The Open University

Longman *London and New York*

Longman Group Limited

Longman House
Burnt Mill, Harlow, Essex CM20 2JE, England
and Associated Companies throughout the world.

Published in the United States of America
by Longman Inc., New York

First published 1982

British Library Cataloguing in Publication Data

Silvertown, Jonathan W.
  Introduction to plant population ecology.
  1 Botany—Ecology
  I. Title
  581.5   QK901
  ISBN 0–582–44265–6

Library of Congress Cataloguing in Publication Data
Silvertown, Jonathan W.
  Introduction to plant population ecology.
  Bibliography: p.
  Includes index.
  1. Plant populations.   2. Vegetation dynamics.
  3. Botany—Ecology.   I. Title.
  QK910.S54   581.5′248   81–15595
                    AACR2

Printed in Great Britain by William Clowes (Beccles) Ltd
Beccles and London

Contents

Author's acknowledgements

As with most books, the name on the cover belongs to only one of the people responsible for the final product. I would like to thank Eva, Alfred, Adrian and the friends who made the task easier by their tolerance of my awkward presence or my awkward absence during its preparation.

A number of colleagues and friends read parts of the manuscript and saved me from blunders. Though I have not accepted all the advice offered, I am very grateful to John Barkham, Steve Bostock, Brian Charlesworth, Deborah Charlesworth, Richard Croucher, Alastair Ewing, Mike Fenner, John Harper, Dave Kelly, Steve Newman, Pat Murphy, Steve Prince, Deborah Rabinowitz, Irene Ridge and Francis Wilkin.

Beverley Simon typed the manuscript with her usual skill and good humour and Niki Koenig drew the plant silhouettes with imagination and care. The staff of the Open University Library relentlessly pursued obscure references and kept me well supplied with scientific papers. I am also grateful to Dennis Baker and to the staff of Longman who guided the book through to publication.

Jonathan W. Silvertown
Open University

This book is dedicated to all students fighting racism

Publisher's acknowledgements

We are grateful to the following for permission to reproduce copyright material:
Academic Press Inc (London) Ltd and the authors for our Fig 3.21 from Fig 2
(Frissell 1973), our Fig 4.6 from Fig 1 (Lovett Doust 1980), our Fig 5.13 from
Figs 1c & d, 2a & d (Wiley & Heath 1969); Acta Forestalia Fennica for our
Fig 6.7 from Fig 12 (Oinonen 1969); The Arnold Arboretum of Harvard
University & Dr A.D. Bell for our Fig 6.10 from Fig 1 (Bell 1974); Bell &
Hyman Ltd for our Fig 4.12 from Fig 3 (Salisbury 1942); Blackwell Scientific
Publications Ltd and the authors for our Fig 6.8 from Fig 1 (Turkington &
Harper 1979), our Fig 2.3a & b from Figs 6a & 7c (Thompson & Grime 1979),
our Fig 2.5 from Figs 7, 8 & 9 (Sarukhan 1974), our Fig 2.8 from Fig 1 (Harper
et al 1965), our Fig 3.8 from Fig 8, 16 & 20 (Sarukhan & Harper 1973), our
Fig 3.9 from Fig 6 (Sarukhan & Gadgil 1974), our Fig 4.5 from left hand side
Fig 7 (Harper & Ogden 1970), our Fig 4.13 from Fig 1.5 (Grime & Jeffrey
1965), our Fig 5.2b from Fig 8 (Watkinson & Harper 1978), our Fig 5.9 from
Fig 4a (Kays & Harper 1974), our Fig 5.10 from Fig 6a, our Fig 5.11 from Fig 1
(Ford 1975), our Fig 5.12 from Fig 4 (Mohler, Marks & Sprugel 1978), our
Fig 6.3 from Fig 2a (Kays & Harper 1974), our Fig 6.1 from Fig 1 (Leith 1960),
our Fig 2.9 from Fig. 3 (Mellanby 1968), our Fig 6.5a from Fig 1, our Fig 6.5b
from Fig 2 (Langer, Ryle & Jewiss 1964); Blackwell Scientific Publications and
University of California Press for our Fig 5.8 from Fig 2.8 Copyright 1980 by
Blackwell Scientific Publications reprinted by permission of the University of
California Press (White 1980); Butterworth & Co Ltd for our Fig 1.4 from left
hand side Fig 14.3 (Fridrikson 1975); University of Colorado for our Fig 2.2
from Fig 3 (Marchand & Roach 1980); CSIRO for our Fig 4.14 from Fig II
(Black 1958), our Fig 7.6 (Hall 1974); Duke University Press & Professor Joan
M Hett from our Fig 5.2a from Fig 1 Copyright 1971 the Ecological Society of
America (Hett 1971); Ekologia Polska for our Fig 2.6 from Fig 7a & b
(Symonides 1977), our Fig 3.5 from Fig 8, 9 (1) (Symonides 1979a), our Fig 5.1
from Fig 18 (Symonides 1979c); J.G.K. Flower-Ellis for our Fig 3.16 from
Fig 31 (Flower-Ellis 1971); Professor J.L. Harper for our Fig 2.4 from Fig 4/1
(Harper 1977); Professor Isao Ikusima for our Fig 1.3 from lower Fig 4
(Ikusima et al 1955); National Research Council of Canada and authors for our
Fig 7.9 from Fig 2 (Turkington et al 1977), our Fig 6.2 from Fig 4 (Thomas &
Dale 1974), our Fig 4.1 from Fig 3 (Eis et al 1965); Nature Conservancy Council
and author for our Fig 8.3 from Fig 3 (Clymo & Reddaway 1972); New
Phytologist for our Fig 2.10 from Fig 1 (Silvertown 1981a); Dr. J. Ogden for
our Fig 4.7 (Harper 1977 after Ogden 1968); OIKOS for our Fig 1.5 from Fig 4
(Callaghan 1976), our Fig 3.11 from Fig 7 (Crisp & Lange 1976); The Open
University for our Fig 4.4 from Fig 26 Open University Unit 2 S364 (c) 1981 The
Open University; PUDOC for our Fig 1.1 adapted from Fig 1 (Harper & White
1971); Dr. A.M. Schaffer for our Fig 4.11 from Fig 22.2 (Schaffer & Schaffer
1977); University of Chicago Press for our Fig 4.9 from Fig 4 (Leverich & Levin
1979), our Fig 8.2 from Fig 2 (Werner & Platt 1976); University of Colorado
for our Fig 4.3 from Fig 3 (Law, Bradshaw & Putnam 1977).

1
Introduction

Why study plant populations?

Two plant ecologists go on a field excursion to a wood. As they walk through the wood the first ecologist begins to make a list of all the plant species present: on the ground there is dog's mercury, wood anemone, wood sanicle and primrose. Among the shrubs and smaller trees hawthorn, dogwood and hazel are recorded and in the tree canopy she records pedunculate oak, ash, birch and field maple plus a note on their relative abundance.

The other ecologist has with her a tree borer, a tape measure and a quadrat. With these she notes down the girth and age of a sample of oaks, the density of oak seedlings and saplings found beneath mature trees and in clearings and some field notes referring to the numbers of seedlings and scattered acorns that seem to have been attacked by animals.

At the end of the day the two ecologists compare notes. The first says: 'This wood falls into the ash – oak wood category as described by Tansley in 1949. It has well-developed herb and shrub layers containing all the species typical of a site on boulder clay.' The other says: 'The dominant species in the wood is *Quercus robur*. Its population is composed predominantly of large trees over 150 years old. There are no younger or sapling oaks but there are many seedlings present beneath the tree canopy and beneath openings at densities up to 100 m^{-2}. The leaves of a high proportion of seedlings beneath the canopy have been eaten by moth larvae. I found a number of scattered remains of acorns that had been eaten by pigeons and squirrels and some that were infested by weevils.' 'The trouble with you,' says the first ecologist, 'is that you can't see the wood for the trees.'

The second woman is a population ecologist. She might reply that only by studying the tree populations of which it is composed and the processes occurring in these populations can the wood really be understood.

Ecology, broadly defined, is the study of interactions between species and their environment. Population ecology is a specialized branch of ecology dealing with the numerical impact of these interactions on a specific set of individuals which occur within a defined geographical area: in other words a *population*. A population ecologist is interested in the numbers of a particular plant or animal to be found in an area and how and why population sizes change (or remain constant). Hence information on the age distribution of trees, the fate of seeds and seedlings and on the predators which affect these potential members of the next generation is vital.

Although it has nineteenth-century antecedents, population ecology, and plant population ecology in particular, is a twentieth-century science. The nineteenth century was the golden age of taxonomy when botanists in Europe and North America spent years of painstaking work classifying plants and naming species. Specimens collected throughout the British Empire were sent to Britain for study. Taxonomists were engaged in a kind of giant stocktake of the biological resources of the British colonies. Kew Gardens was an important centre for this work (Brockway 1980). Via Kew rubber (*Hevea brasiliensis*) was introduced into South-East Asia from South America. Holland, France and Britain all introduced coffee (*Coffea arabica*) into their colonies. Like rubber, the main centre for coffee cultivation is now in non-indigenous areas (Purseglove 1968).

Plant ecologists working at the beginning of the twentieth-century were strongly influenced by the nineteenth-century preoccupation with classification. Hence a good deal of time was spent classifying and naming plant communities, almost as if they were organisms in their own right. For a while an argument even raged about whether plant communities developed through fixed sequences of species, analogous to the development of individual organisms through a sequence of embryonic stages (Phillips 1934). Although the integral view of plant communities was challenged soon after it emerged (Gleason 1926, 1927), the emphasis on community classification has remained strong in plant ecology into the 1970s.

During this period applied ecologists studying forests or field crops saw things differently. They were (and are) interested in the *number* of seeds that must be sown to obtain a crop with an economic *yield*; the *density* at which seeds should be sown; the effects of varying *proportions* in mixtures of species; the causes of mortality and the *quantitative effects* of competitors (weeds), predators (insect pests, etc.) and disease. These studies form the historical foundations of plant population ecology.

The demography of plants

Stages in the plant life cycle provide useful intervals at which to analyse changes that take place in plant population size with time. Demography is the study of these population changes and their causes throughout the life cycle. The basic stages of a typical plant life cycle are illustrated diagrammatically in Fig. 1.1. This diagram is self-explanatory, but there is some conventional terminology applied to various stages depicted in it.

The seed population in the soil is generally referred to as a *seed bank* or a *seed pool*; the latter term is used in this book. The interface between the seed pool and the establishment of seedlings is often envisaged as an *environmental sieve*. Some seeds pass through it to successful seedling emergence, others die or remain dormant in the soil. Seedlings which emerge simultaneously or nearly so form a *cohort*. The cohort is a particularly important unit in demography. By following the fate of a cohort of individuals through time, we can obtain values of probability for giving birth (seed production), death and survival for typical individuals of specific age.

The transition from juvenile stages such as seeds or seedlings to later stages in the life cycle in which reproduction occurs is called *recruitment*. Recruitment may occur from seeds or, as is the case in many plants, by the production of *vegetative daughters*. These daughters are initially physically attached to the parent, which may itself be no more than a collection of rosettes, stems or tillers (in grasses) each capable of an independent existence if they are detached. These morphological units with the potential for an independent existence are called *ramets*. Vegetative daughters or ramets which have all been produced from the

Fig. 1.1. An idealized plant life history. (Adapted from Harper and White 1971)

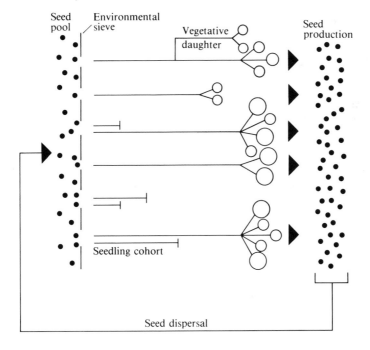

same parent constitute a *clone*. A plant, of whatever size and however divided into ramets, which originates from seed is called a *genet*, as all parts share exactly the same genes.

Virtually all plants possess a modular structure which consists of a multiplication of basic units or modules. A typical module is an aerial shoot with lateral meristems (from which other shoots may arise) and terminating in an inflorescence. The region between two nodes of a stem or a rhizome (an internode) may also form a module of plant structure. Collections of aerial modules produced by repeated branching form plants of increasing size. The number of branches is an important determinant of how large a plant is and how many leaves, flowers and fruit it carries. Repeated branching and addition of horizontal modules, such as those of which rhizomes are composed, may extend the area over which a plant spreads and will affect the number of vegetative daughters produced. Thus the structure of individual plants and the population structure of clonal plants are both expressions of the modular architecture of plants (J. White 1979).

Population dynamics and evolution by natural selection

Most of this book is concerned with changes in the numbers of plants and with changes in the relative proportions of different species in populations. At a finer level of analysis we would find that most single-species populations consist of a collection of individuals which differ to some degree in characteristics such as leaf shape, flower colour or in biochemical properties and other aspects of their outward appearance or *phenotype*. Differences in genetic make-up (genotype) often underlie phenotypic differences between individuals. Only the phenotype is visible to us but this is the 'public face' of the genotype, some of whose properties may be deduced by growing different individuals in the same environment in transplant experiments, or from breeding experiments with different phenotypes.

For instance in populations of white clover (*Trifolium repens*) two distinct types of plant are found. When leaves are damaged, one type of plant produces free cyanide which is thought to deter slugs and other mollusc predators (Crawford-Sidebotham 1972). The other type of plant does not produce the cyanide poison. Thus there is a cyanogenic and a non-cyanogenic phenotype.

Breeding experiments have shown that two genetic loci determine cyanogenesis. One locus controls the production of cyanogenic glucosides and the other controls the production of an enzyme which breaks down the glucosides to liberate cyanide. The cyanogenic phenotype occurs only when the genotype has the correct combination of alleles for production of both the enzyme and the glucoside. There are three genotypes which have acyanogenic phenotypes: 1. the genotype lacking the glucoside allele; 2. the one lacking the enzyme allele; and 3. the genotype lacking both.

Evolutionary changes can occur in a population if three conditions are met: 1. there is phenotypic variation; 2. some of this variation is heritable (genetic); and 3. selection acts upon different phenotypes. The potential for evolutionary change in plant populations is illustrated in any greengrocers' shop where cabbage, cauliflower, broccoli, Brussels sprouts and kohlrabi are on display. All these vegetables and their many varieties have been derived by artificial selection from the same ancestral species of wild cabbage (*Brassica oleracea*). Generations of artificial selection have transformed this apparently unappetizing and unpromising plant into varieties with a greatly enlarged terminal leaf bud (cabbages), a proliferation of axillary buds (Brussels sprouts), a large, swollen inflorescence of undeveloped flowers (cauliflower), several lax, terminal inflorescences (broccoli) or a swollen, bulbous stem (kohlrabi).

Natural selection may produce results just as dramatic as those of artificial selection but is less rapid. It occurs when one phenotype leaves more descendants than another because of its superior ability to survive, to produce offspring or due to a superiority in both of these characters. Notice that *survival* and *reproduction* are both demographic processes, and hence natural selection is also a demographic process. Where it is possible to analyse the demography of different phenotypes in a population separately, it is also possible to determine which phenotype is likely to leave the most descendants and hence in which direction(s) natural selection is acting.

Survival and reproductive success are combined into a single measure of relative evolutionary advantage called *fitness*. The fitness of a particular phenotype is not a fixed value, but is determined in the context of prevailing ecological conditions and relative to the survival and reproductive success of other phenotypes which occur in the same population.

For example, cyanogenic *Trifolium* may have a higher fitness than acyanogenic *Trifolium* in the presence of slugs but when slugs are absent the relative values of fitness for the two phenotypes may be reversed. Cyanogenic plants are more prone to frost damage than acyanogenic plants, and without slugs to redress this disadvantage or in particularly cold areas, the latter phenotype may survive and reproduce the most successfully.

Although population dynamics lays the basis for an assessment of selective forces in natural populations, this absorbing subject is mostly beyond the scope of this book. However, the relevance of natural selection in plant populations is briefly touched upon in most chapters, particularly in Chapter 4. We must now address the most basic subject in population ecology: population growth.

Models of population growth

The simplest model of population growth, and the most intuitively obvious one, is based upon the straightforward quantities of three properties of population change; births B, deaths D and migration (E for emigration, I for immigration). The magnitude of these population properties, known as population *parameters*, allows us to state the size of a population N_{t+1} after one generation has elapsed since its size was N_t.

$$N_{t+1} = N_t + B - D + I - E \qquad [1.1]$$

The net rate at which the population is increasing or decreasing is called the *net reproductive rate*, R_o, and is simply the ratio of offspring produced in generation N_{t+1} to the population size in the previous generation N_t:

$$R_o = N_{t+1}/N_t \qquad [1.2]$$

It is worth returning to equation [1.1] which must be the nearest thing there is to an algebraic truism, whenever the going gets tough with more complex descriptions of population change. However, the mathematics in this book are kept to a minimum and should give the reader little cause to resort to this comforting equation. All equations for population change more complex than [1.1] arise from attempts to account for the way in which birth-*rates*, death-*rates* and migration *rates* alter with population density and age structure and with the effects of competitors, predators, pathogens and mutualists. A thorough mathematical treatment of these relationships is outside the scope of this book, but we will deal with the simplest models of the effects of density and age structure in this chapter.

When birth-rates b and death-rates d in a population are constant we may calculate the *instantaneous rate* of increase r from these two parameters:

$$r = b - d \qquad [1.3]$$

The rate of change of population size dN/dt is then given by the differential equation:

$$\frac{dN}{dt} = rN \qquad [1.4]$$

where N is the size of the population at a particular instant in time. To obtain the size of the population N_t at some time t, [1.4] is integrated to give:

$$N_t = N_0\, e^{rt} \qquad [1.5]$$

Where N_0 is the initial size of the population at some time designated zero, e is a constant $= 2.718$ (the base of natural logarithms), r is the intrinsic rate of increase as defined in [1.3] and t is the time elapsed since time zero. The population growth described by [1.5] is *geometric*, which means that population size goes on doubling (or halving if $d > b$) at a constant rate as shown in Fig. 1.2, curve a. This situation is clearly unrealistic, since the growth of all plant populations must be limited by the availability of some resource, even if it is only the total area of land surface on earth! Equation [1.4] can be modified to take account of the limitations to population growth which operate as population size increases towards the limits of resource availability, if we add a term which reduces rN as that limit, called the carrying capacity K, is reached:

$$\frac{dN}{dt} = rN \frac{(K - N)}{K} \qquad [1.6]$$

This is the *logistic* equation and its behaviour is shown in Fig. 1.2, curve b. It approximately describes the growth of laboratory cultures of duckweed (Fig. 1.3) and the colonization of new rock surface by mosses on the volcanic island of Surtsey (Fig. 1.4), perhaps two of the simplest situations an ecologist could ever hope to study.

The geometric and logistic models we have just considered are based upon the assumption that population growth is continuous and they employ an *instantaneous* rate of increase r. Although this approach is of some descriptive value, it is not particularly realistic for most plant populations in which reproduction is usually confined to one part of the year and to plants which have reached a minimum age or size. Simple *matrix* methods allow us to model population changes where individuals fall into different age or size classes and have different rates of reproduction and death at different ages. These models give some insights into population growth which cannot be obtained from continuous models of growth.

First, the number of individuals present in each age class in the particular population under study is entered in a *column matrix*. The

Fig. 1.2 (a) Geometric; and (b) logistic population growth.

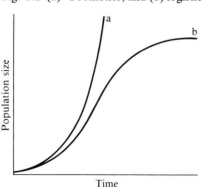

number of age classes employed depends upon the type of life history of the population being studied and how often it is censused. If we are studying an annual plant by a once-yearly census at flowering time then there are effectively only three age classes in the population: seeds, non-flowering rosettes and flowering individuals. If we call these age classes N_s for the number of seeds, N_r for the number of rosettes and N_f for the number of flowering individuals in a particular generation, the column matrix is:

$$\begin{bmatrix} N_s \\ N_r \\ N_f \end{bmatrix}$$
Matrix 1.
A column matrix of three age classes

The object of modelling population change is to be able to predict the magnitude of N_s, N_r and N_f in the following year, and to do this we must know how many seeds grow into non-flowering rosettes, how many grow into flowering individuals and how many seeds are produced (on average) by each flowering individual. This information is obtained by the demographic methods discussed in greater detail in Chapter 2. The results of the demographic study of our model annual plant should provide us with values (coefficients) for the probability of a seed making the transition to later stages of the life history, hence these coefficients are known as *transition probabilities* and they have values between 0 and 1. These are entered in a *transition matrix*. This matrix is square, with each side the same length as the *column matrix*, so that the transition matrix can accommodate coefficients for the probability of every transi-

Fig. 1.3 Growth of an experimental population of duckweed *Spirodela oligorrhiza*. (From Ikusima *et al.* 1955)

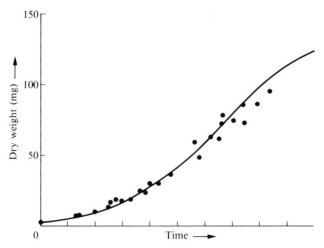

tion from one age class to another:

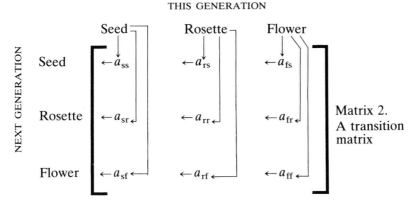

In matrix 2, a_{ss} is the probability that a seed this year will not germinate but will remain a viable seed next year, a_{sr} is the probability that the seed will become a rosette next year and a_{sf} is its probability of flowering and so on. The first column gives the fate of seeds, the second the fate of rosettes and the third the fate of flowering plants. Some of the coefficients in the transition matrix have no biological meaning, for

Fig. 1.4 Growth of a moss population colonizing bare rock on the Icelandic island of Surtsey. (From Fridrikson 1975)

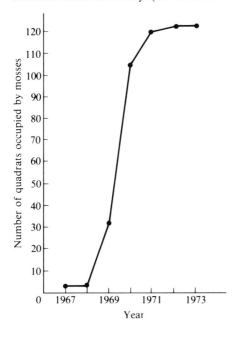

instance by definition non-flowering rosettes cannot produce seeds and a_{rs} is replaced by a zero when we write out the values of the matrix. Other coefficients in the matrix may be zero when certain transitions do not occur because of the peculiarities of the life history of the population. If we imagine our model *annual* plant behaves true to its description, then all rosettes which have not flowered in their first year will die and no flowering individuals will live after setting seed. In this case a_{rr}, a_{rs} and a_{fr} are all zero and the transition matrix is:

$$\begin{bmatrix} a_{ss} & 0 & a_{fs} \\ a_{sr} & 0 & 0 \\ a_{sf} & 0 & a_{ff} \end{bmatrix}$$

Matrix 3.
A transition matrix for an annual

Notice in matrix 3 that a_{ff}, the transition probability for flowering individuals producing other flowering individuals, is not zero. This transition is in fact the probability for the route flower → seed → flower, but since this takes place within a year in our model population and we have only censused twice in this period (at flowering time) we only have information on the overall probability: flower → flower. Another census in the middle of the year would provide more information about the population and allow us to construct two transition matrices, one for events in the first half of the year and another for the second. These complications of the method will be ignored for the moment.

If we are dealing with a plant such as creeping buttercup *Ranunculus repens* which is longer lived and produces rosettes by vegetative propagation, the transition matrix for a yearly census will look slightly different:

$$\begin{bmatrix} a_{ss} & 0 & a_{fs} \\ a_{sr} & a_{rr} & a_{fr} \\ 0 & a_{rf} & a_{ff} \end{bmatrix}$$

Matrix 4.
A transition matrix for a perennial which produces vegetative daughters

Many seeds remain dormant in the soil from one year to the next and thus $a_{ss} \gg 0$; a very few rosettes arise from seed each year, i.e. $a_{sr} \gg 0$, but no seeds reach flowering within a single year so $a_{sf} = 0$ and a large proportion of flowering individuals and rosettes produce vegetative daughters, i.e. $a_{fr} \gg 0$ and $a_{rr} \gg 0$.

The transition matrix is given the conventional notation A and the column matrix B_t where t is the generation (or other time interval) at which the age structure of the population is determined. Matrix multiplication of B_1 by A gives a new column matrix B_2 describing the population in the next generation. Starting with known values for the coefficients in A and some known quantities for B_1, we obtain the new value of N_t in B_2 by multiplying each coefficient (a_{ss}, 0, a_{fs}) in the first row of A by the corresponding value (N_s, N_r, N_f) in B_1 and then summing the products. The new value of N_r is obtained by multiplying each coefficient $(a_{sr}, 0, 0)$ in the second row of A by the corresponding value (N_s, N_r, N_f) in B_1 and summing the products and so on. It may be useful at this point to refer back to the explanatory diagram of a transition matrix (matrix 2) for the biological meaning of the multiplication $A \times B_1$. The algebraic result is as follows:

$$
\begin{array}{ccc}
A & \times \quad B_1 = & B_2 \\
\text{Matrix} \quad \begin{bmatrix} a_{ss} & 0 & a_{fs} \\ a_{sr} & 0 & 0 \\ a_{sf} & 0 & a_{ff} \end{bmatrix} & \begin{bmatrix} N_s \\ N_r \\ N_f \end{bmatrix} & \begin{bmatrix} (N_s\, a_{ss}) + (N_f\, a_{fs}) \\ (N_s\, a_{sr}) \\ (N_s\, a_{sf}) + (N_f\, a_{ff}) \end{bmatrix}
\end{array}
$$

A numerical example of the matrix operations just described is given in the Box. When any transition matrix is iterated (i.e. repeatedly multiplied; $A \times B_1 = B_2$, $A \times B_2 = B_3$, $A \times B_3 = B_4$, etc.) the age structure of the population eventually stabilizes at a constant ratio of age classes (e.g. seeds: rosettes: flowering plants) which depends upon the values of the coefficients in A and is independent of the values in B_1. This is demonstrated by the example in the Box. It has also been proved as a general theorem applying to all matrices of this kind when A contains coefficients which remain constant. Once $A \times B_t$ has been iterated to the point at which a stable age distribution is reached, the ratio of any one age class (say N_s) to the same age class in the next generation gives us R_o, or the *net rate of reproduction* for the population.

The deliberate emphasis in this section has been on matrices as a tool rather than on the biological meaning and experimental derivation of transition probabilities and age structures. The great advantage of the matrix method is that it allows us to determine the effect of changing transition probabilities such as those between the stages seed → rosette or between flowering plant → seed. This matrix tool is referred to on several occasions in the following chapters and it is hoped that this preliminary account of the simple methods involved will prevent any confusion in the reader's mind between the inherent properties of the method (such as the division of the year or generation into census periods or the stable age distributions produced when transition probabilities are constants), and the properties of plant populations themselves. Natural populations undergo continuous periods of change governed by transition probabilities which may alter with changing ecological conditions. It is these properties of plant populations which are the real subject of this book.

Predicting population changes by the matrix method: Carex bigelowii

A population of the sedge *Carex bigelowii* growing in a lichen heath in alpine tundra in Norway was studied by Callaghan (1976). The plant has an extensive underground rhizome which produces tillers (aerial shoots) at intervals along its length as it grows. Each of these tillers may flower and die or initiate new lateral shoots of one or more daughter tillers. The sequential addition of new tillers along these rhizomes, whose growth is directional, and the persistence of dead tillers *in situ* on it makes it possible to reconstruct the history of a tiller population. Callaghan divided the 320 living and dead tillers he found on a sample of 23 intact rhizome systems into 4 growth stages according to the number of leaves on them. (Callaghan himself calls these 'age classes' but they actually have no definite known timespan and so are better described as size classes or stages.) From this information he derived a stage structure for the total tiller population. These data, in units of numbers of tillers per square metre, provide the entries in the column matrix B. The bottom row of the transition matrix A was filled out with the probabilities of plants dying, calculated from the ratio of dead to dead + living tillers in each stage. For example 55 per cent of all first stage tillers counted were dead so the assumed probability of death in this stage is 0.55. The 45 per cent of surviving tillers in the first stage are assumed eventually to enter the next stage so the transition $1 \rightarrow 2$ is given the value 0.45. Other transitions between successive stages were calculated in the same way. The addition

Fig. 1.5 Changes in stage class structure predicted for the *Carex bigelowii* population from transition matrix A. (From Callaghan 1976)

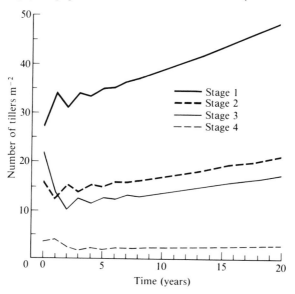

of new vegetative daughters to stage 1 from tillers in all stages was calculated from counts of the number of new lateral tillers associated with each tiller on the main axis of the rhizome. Tillers flowering in stages 2–4 produced seed but no seedlings of *C. bigelowii* were observed at Callaghan's site and he therefore assumed no recruitment to the population from seed. Seedlings of this species have been observed in studies of the plant in Iceland and must occur occasionally in Norway too. Omitting them from the present model is an admitted drawback to Callaghan's method of deriving a transition matrix for this species. Nevertheless, using the derived values in matrices *A* and *B* and iterating the model through twenty cycles, the population is predicted to increase steadily (Fig. 1.5). Notice that after some initial fluctuations the number of tillers in each stage begin to approach a constant proportion of the total population. This is a property of matrix models which employ unchanging transition probabilities. In reality we might expect the values in *A* to change as a rhizome system senesced. One effect of senescence could be a reduction in the number of vegetative daughters produced per tiller. Readers may calculate for themselves what effect such a reduction would have on the growth of the tiller population.

STAGE	1	2	3	4	dead		Number/m^2	STAGE
1	0.39	1.39	0.04	0	—		27.18	1
2	0.45	—	—	—	—		16.14	2
3	—	0.83	—	—	—		22.08	3
4	—	—	0.18	—	—		3.40	4
dead	0.55	0.17	0.83	1	1		—	dead

Transition matrix			Column matrix
	A		*B*

Summary

Plant communities have traditionally been described and classified in a manner which ignores the dynamic nature of the populations of which they are composed. Population ecology is principally concerned with the processes which determine population size and population changes. Historically, this approach owes a great deal to the work of applied ecologists.

Demography is the quantitative study of population changes throughout the life cycle. The phases of a model plant life cycle are the *seed pool*, the *environmental sieve*, the seedling *cohort* and varying stages of juvenile development leading up to reproductive maturity. New plants may be *recruited* to the population from *vegetative daughters* or from seed. *Clonal* plants rely mainly on vegetative propagation. A plant originating from seed is a *genet*. Individual plants and clones both have a *modular structure*.

Fitness is a relative measure of evolutionary advantage which is based upon the survival and reproductive success of different phenotypes. *Natural selection* is a demographic process.

Two simple models of population growth are the equation for *exponential* increase and the *logistic equation*. Populations with an *age structure* may be modelled using *matrices*.

2
Life tables and some of their components

A newspaper which specializes in titillating its Sunday readers with gossip and scandal proclaims on its masthead that 'All human life is here'. A somewhat naïve demographer, opening a paper with this masthead, would expect to find the contents crammed with life tables and fecundity schedules, for as far as demography is concerned, in these all life may be found. Life tables and fecundity schedules summarize all the most important events in a population: the births, the deaths and – essential information to gossip writers and demographers alike – the *age* of the individuals who are dying or giving birth.

Life tables and fecundity schedules

Life tables were first drawn up by actuaries who needed a precise set of data on mortality in the human population in order to be able to assess the insurance risk that is attached to different individuals. Although people may die at any age and from a variety of causes, statistically (i.e. on average) the death risk to an individual is related to that individual's age. Life tables therefore divide the population into age classes, each of which has an *age-specific mortality risk*.

The simplest way of compiling a life table is to follow the fate of the individuals in a cohort from birth until the last member of the cohort dies. This 'following' procedure produces a *dynamic life table*. Another method commonly used in situations where it is not practical to follow the demise of a cohort through time (e.g. for long-lived trees) estimates age-specific death risks from the age structure of a population at one moment in time. This produces a *static life table* and is discussed in Chapter 3.

An example of a dynamic life table for an annual plant, *Phlox drummondii*, is shown in Table 2.1(a). Populations of this plant were studied in Texas, USA, by Leverich and Levin (1979). Most plant populations contain overlapping generations, thus complicating the methods required to obtain a cohort of uniform age. This is particularly a problem in the seed fraction of plant populations, since dormant seeds often accumulate in the soil, year after year. Viable seeds of *P. drummondii* apparently do not persist in the soil beyond one season so no overlap of generations occurred in the seed fraction of this

population. Furthermore, seeds germinated approximately synchro-
nously, so that all members of the population effectively belonged
to the same cohort and formed a *discrete generation*. Both of these
factors simplified the collection of data considerably.

Table 2.1(a) Life table for *Phlox drummondii* at Nixon, Texas

Age interval (days) $x - x'$	Length of interval (days) D_x	No. surviving to day x N_x	Survivorship l_x	No. dying during interval d_x	Average mortality rate per day q_x
0– 63	63	996	1.0000	328	0.0052
63–124	61	668	0.6707	373	0.0092
124–184	60	295	0.2962	105	0.0059
184–215	31	190	0.1908	14	0.0024
215–231	16	176	0.1767	2	0.0007
231–247	16	174	0.1747	1	0.0004
247–264	17	173	0.1737	1	0.0003
264–271	7	172	0.1727	2	0.0017
271–278	7	170	0.1707	3	0.0025
278–285	7	167	0.1677	2	0.0017
285–292	7	165	0.1657	6	0.0052
292–299	7	159	0.1596	1	0.0009
299–306	7	158	0.1586	4	0.0036
306–313	7	154	0.1546	3	0.0028
313–320	7	151	0.1516	4	0.0038
320–327	7	147	0.1476	11	0.0107
327–334	7	136	0.1365	31	0.0325
334–341	7	105	0.1054	31	0.0422
341–348	7	74	0.0743	52	0.1004
348–355	7	22	0.0221	22	0.1428
355–362	7	0	0.0000		

From Leverich and Levin 1979

Table 2.1(b) Fecundity schedule for *Phlox drummondii* at Nixon, Texas

$x - x'$	B_x^{seed}	N_x	b_x^{seed}	l_x	$l_x b_x$
0–299	0.000	996	0.0000	1.0000	0.0000
299–306	52.954	158	0.3394	0.1586	0.0532
306–313	122.630	154	0.7963	0.1546	0.1231
313–320	362.317	151	2.3995	0.1516	0.3638
320–327	457.077	147	3.1904	0.1476	0.4589
327–334	345.594	136	2.5411	0.1365	0.3470
334–341	331.659	105	3.1589	0.1054	0.3330
341–348	641.023	74	8.6625	0.0743	0.6436
348–355	94.760	22	4.3072	0.0221	0.0951
355–362	0.000	0	0.0000	0.0000	0.0000
					$\Sigma = 2.4177$

From Leverich and Levin 1979

A census of the seed population was carried out seven times before germination and then at seven-day intervals until all remaining individuals flowered and died. These intervals are indicated in the first column of the life table. The second column records the length, in days, between two successive censuses, the third column records the number of survivors N_x present at the beginning of an age interval and the fourth column the proportion l_x surviving to day x. Life tables are conventionally drawn up with 1000 individuals in the cohort on the first census (day zero). Of course the actual number of individuals counted may be more or less than 1000 and N_x values may have to be adjusted to allow for this.

The average mortality rate per day q_x is obtained by dividing the number dying d_x during an interval by the length of the interval D_x.

Largely because actuaries were only interested in the relation between survival and age and had no use for information on births, fecundity data are traditionally recorded separately from the life table but in a parallel fashion. The mean number of seeds produced by individuals (or females of dioecious species) in an age interval $x - x'$ is given the symbol b_x. This information and other measurements of seed production derived from it are shown for *P. drummondii* in the fecundity schedule in Table 2.1(b).

Life tables and fecundity schedules contain the data required to model the dynamics of populations using the matrix method. Death-rates and birth-rates derived from these tables become the transition probabilities of a matrix model. Although the data collected in a dynamic life table must come from a cohort of individuals of uniform age, when the age-specific rates of death and birth are plugged into a probability matrix we may generate a model population which contains overlapping generations. Some examples of these models are considered in Chapter 3.

In species where growth is determinate and the probabilities of survival and giving birth are closely related to age, it makes elementary sense to tabulate birth and death in terms of the age of individuals. Actuaries would lose a good deal of accuracy in their calculations if they based their estimate of someone's chances of survival on some parameter only loosely related to it such as body-weight. Age is not necessarily the best predictor of the fate of an individual in all species, and in particular in plants where growth is extremely plastic we may expect to find other parameters which are more useful.

Age versus stage

The practice of plant demographers in the Soviet Union since the 1940s has been to classify individuals in a population according to their *stage* of growth and reproduction rather than according to their chronological age. T. A. Rabotnov and other Soviet ecologists have used this method in a large number of demographic studies, several of which have followed the fate of cohorts of herbaceous plants over periods of 10 years or more.

In one such study of the perennial *Ranunculus acris* growing in a floodplain meadow (Rabotnov 1964, 1978a) individuals in a 10 m^2 plot were classified into the categories juvenile (J), immature (I), vegetative (V), generative (flowering, G) or dead (D). When the first records were made in 1950 there were 178J, 125I, 123V and 25G. The fate of these plants was recorded the following year and is shown in Fig. 2.1(a). Nearly equal proportions of the 178 juveniles died or grew to the I stage, and small proportions remained J or became V. The 1950 cohort of immatures also had mixed fortunes, with as many plants remaining in the I stage as transferring to the V stage. Small proportions of the 1950 Vs and Gs actually regressed to earlier stages in 1951, but the majority in each cohort remained in the same stage of growth in which they were found during 1950. Only a very small proportion of the 1950 Vs and Gs died in 1951. It is clear from this study that population changes in *R. acris* are complex and that the fate of an individual plant is not strictly determined by its chronological age. In fact these kinds of transitions between one growth stage and another cannot be described in a life table which is designed for populations in which all individuals move inexorably from one age interval to the next.

Plant populations appear to be governed by more probabilistic processes than those which determine the fate of individuals in the populations for which life tables were originally designed. Unfortunately we cannot place exact chronological ages on plants in the I, V or G stages to judge the importance of age in the *R. acris* study properly because some plants originated before 1950 when the study began. Nevertheless we know that *R. acris* does not escape the penalties of old age entirely because transition probabilities between growth stages did alter as the study progressed. Figure 2.1b shows the fate in 1953–54 of 276 survivors from the original plants of 1950. Death rates in all three remaining stages rose dramatically: from 0.22 to 0.65 among I's and from 0.08 to about 0.44 in V's and G's.

The relative importance of age versus stage of plant growth in determining the fate of individuals has been compared in experimental populations of teasel (*Dipsacus fullonum*) by Werner (1975) and Werner and Caswell (1977). This plant is generally referred to as biennial, but like many plants described in this way it actually often takes more than 2 years for an individual to reach flowering, which is always followed by death. Werner overcame the problem of determining the age of plants by sowing populations in field sites where *D. fullonum* was previously absent and following their fate from seedling emergence. Individuals were mapped over a 5-year period and rosette size and vegetative or flowering condition were noted. Rosettes were divided into different size classes and it was found that the probability of a rosette flowering was strongly correlated with its size but independent of its age. In contrast to the behaviour of *R. acris*, transition

Fig. 2.1 (a) The fate, 1 year later, of juvenile (J), immature (I), vegetative (V), generative (G) individuals of *Ranunculus acris* from a population marked in 1950; (b) the fate of survivors of the 1950 population between 1953 and 1954. Numbers in *italics* are the sample sizes for each growth stage. The width of arrows is proportional to the probability of the transitions between stages. (Data from Rabotnov 1978a)

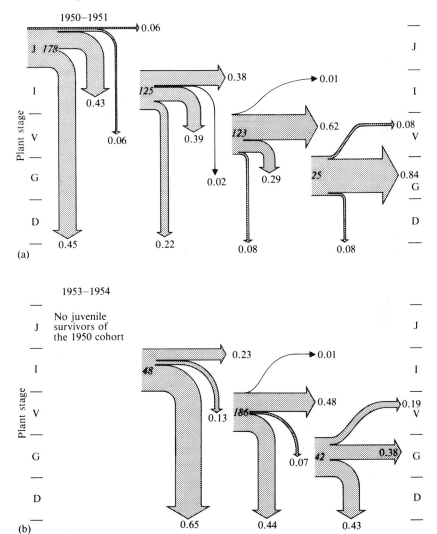

probabilities for rosettes at a given stage of growth (size) were appreciably the same whether the rosettes were 2, 3, or 4 years old.

Werner and Caswell (1977) compared the accuracy of predictions of population size made by matrix models based upon age-related transition probabilities and stage-related transition probabilities. They found that the stage-related models predicted changes in the number of seeds and vegetative or flowering rosettes found in experimental populations better than the age-related models, even though the transition probabilities for both types of models were derived from the same populations. The lower accuracy of the age-related models was evidently due to the fact that events in the life history of teasel are not as closely tied to age as they are to the growth stage of these plants.

Flowering behaviour is related to rosette size in a number of monocarpic perennials (e.g. *Pastinaca sativa*, Baskin and Baskin 1979). The size of a rosette in these plants is usually an indication of the size of the tap-root beneath, in which carbohydrates are stored. The tap-root supplies these carbohydrates to the above-ground part of the plant when flowering occurs (Glier and Caruso 1973), so the size of the tap-root itself would probably provide a more accurate prediction of flowering. Unfortunately it cannot be measured easily, without disturbing the plant.

The size of the storage organ is more easily measured in species which possess a bulb. The age and size of bulbs are closely related to each other in most species with such an organ. This makes the independent effects of age and size on flowering difficult to assess. Nevertheless flowering appears to be related to bulb size, independently of age, in commercial tulips (Fortainier 1973), and probably also in wild daffodil (*Narcissus pseudonarcissus*) (Barkham 1980) and a number of other bulbiferous woodland herbs (Kawano 1975; Kawano and Nagai 1975).

Dormant buds, oskars and life below ground

Analysing plant populations in terms of stages makes even more sense when we examine the recruitment phase of plant life histories. Plants exhibit a number of devices by which individuals persist for years in a dormant state awaiting a period of amelioration in the environment before they embark on full growth and the road to reproduction. The most common form of this waiting game is played by dormant seeds, but dormant buds on rhizomes play the same role in couch grass (*Agropyron repens*) (Tripathi and Harper 1973) and sand sedge (*Carex arenaria*) (Noble, Bell and Harper 1979).

With some exceptions, trees do not possess dormant seeds in the soil. Instead, many tree seedlings may be found in populations of ageing juveniles, lingering in a stunted condition in the field layer, far beneath the tree canopy. We could describe this habit as the *Oskar* syndrome, after the character in Günther Grass' novel *The Tin Drum*, who preferred the juvenile to the adult state and stopped growing at the age of three. Oskars occur in a diverse variety of species including hemlock (*Tsuga canadensis*), Norway spruce (*Picea abies*), white oak (*Quercus alba*), holly (*Ilex aquifolium*) and beech (*Fagus grandifolia*) (Grime 1979).

The striped maple *Acer pensylvanicum* which occurs in the eastern USA has oskars on the forest floor where it occurs as an understorey tree. These oskars persist for up to 20 years suffering little mortality while they await an opening in the tree canopy. If an opening appears, they grow rapidly to fill it, flower and reproduce (Hibbs and Fischer 1979).

An analogous strategy is found in Prescott chervil (*Chaerophyllum prescottii*) which is a monocarpic umbellifer found in meadows in the forest steppes of the USSR. This plant has no detectable reserve of dormant seed in the soil but possesses a dormant underground tuber which may vary from the size of a pea to that of a hen's egg. These tubers are produced from the transformed tap-roots of vegetative rosettes whose leaves and outer roots shrivel and die after several seasons' growth. The tuber left behind by this vegetative stage may remain dormant in the soil for over 10 years and it does not enter the next stage of growth until the surrounding vegetation is disturbed. When meadows are ploughed or areas of grass are killed under a haystack, hundreds of dormant chervil tubers are suddenly activated into growth. The tubers produce a system of surface roots, leaves are formed and flowering takes place, followed by death. Seeds germinate immediately they reach the soil, allowing seedlings to take advantage of the area of bare ground created by the same disturbance which originally stimulated flowering (Rabotnov 1964).

There are a number of examples of this kind of interrupted life history among terrestrial orchids. Orchid seeds are among the smallest produced by any angiosperm; they are so small that probably no auto-trophic seedling could establish itself from such an impoverished beginning. The usual pattern of development, following the germination of an orchid seed, involves a more or less prolonged period during which the orchid depends upon a fungus for a supply of nutrients. During this parasitic phase of life, a cigar-shaped mycorrhizome, with no leaves or aerial parts, is formed. After subterranean growth of some years, the enlarged mycorrhizome produces an aerial shoot. Shortly afterwards the mycorrhizome itself disappears and is replaced by one or more tubers which contain no fungal hyphae. Underground development may be protracted, for instance the burnt orchid (*Orchis ustulata*) may take up to 15 years to produce its first aerial shoots and flowers. Thereafter, most orchids perennate by the production of new tubers and new shoots each year.

In some orchids, plants which have already reached the flowering stage, and which have lost their mycorrhizome, may regress, developing a new mycorrhizome and returning to an exclusively underground existence. Red helleborine (*Cephalanthra rubra*), lady's slipper (*Cyprepedium calceolus*) and creeping lady's tresses (*Goodyera repens*) are woodland species which behave in this manner when the tree canopy in their habitat closes, excluding light. Plants of red helleborine have been known to reappear after 20 years of subterranean life, when the tree canopy has opened again (Summerhayes 1968).

Certain vines of evergreen tropical rainforest exhibit seedling behaviour which incorporates both the Oskar syndrome and an underground storage organ. A small understorey shrub is produced which remains in a stunted condition for a considerable period of time. While in this state, the oskar produces a large tuber from its tap-root. Then, quite suddenly, the central stem of the shrub begins to elongate. Growing as much as 5 cm per day, it may reach more than 5 m into the tree canopy before it produces the leaves and clambering stems of a mature vine. Janzen, who describes the habit of these plants, suggests that the main advantage of this sudden switch from inactivity to rapid growth may be that it exposes the tender growing tip of the climbing stem to predators for only as short a time as possible. A stem elongating more slowly and steadily would probably be more vulnerable. The tuber stores the resources which make a sprint for the tree canopy possible in these vines. The tuber itself is therefore potentially a rich, concentrated food source for predators. In fact it is less vulnerable than other parts of the plant because tubers are usually protected by large quantities of toxic secondary compounds (Janzen 1975a).

Seed dispersal

Plants are very poorly mobile by comparison with animals, for even sedentary animals such as barnacles have highly mobile larval stages. It is perhaps this very lack of migratory ability in plants which has reinforced the evolution of mechanisms which allow plants to persist through unfavourable periods in such a variety of ways.

Seeds are, of course, the main means of dispersal for higher plants and their movement is of interest to the population ecologist for two reasons: firstly, seeds may augment or deplete local populations, thus affecting population size and secondly, small numbers of dispersing seeds may act as founders of new populations which may grow to significant proportions within a few generations. Pollen, which can move much greater distances than seeds, is a significant transporter of genes between established populations but cannot produce either of the ecological results of seed movement. The effects of both seed and pollen movement on gene flow are beyond the scope of this book but are comprehensively reviewed by Levin and Kerster (1974).

Seeds dispersed by wind, whether aided by a flight appendage such as a wing or a pappus or not, generally move only short distances from the mother plant (Sheldon and Burrows 1973). Figure 2.2 shows the distribution of seeds with distance from plants of four species of arctic alpine herbs which colonize gaps in tundra vegetation. Although these plants are all colonists of ephemeral sites, their dispersal ability is not good. The shape of the dispersal curves for these species typifies patterns of seed fall found in a variety of other herbs, many of which deposit most of their seeds near the base of the plant, or at a short distance from it.

The seeds of trees have further to fall than those of herbs and are consequently carried further from their point of release. Nevertheless, the vast majority of tree seeds are deposited within tens of metres of the parent and dispersal curves are similar in shape to those of herbs.

Animal dispersal of seeds is a much more erratic process than wind dispersal. Except for those cases where birds or rodents cache seeds, it seems unlikely that seeds are moved in large numbers to any one place. On the other hand the role of animals, particularly birds, in transporting occasional founders to new sites of colonization is probably very significant.

Recruitment from buried seeds

The production of seeds which become incorporated into a buried population is such a common feature of plant life histories that a seed pool can be found beneath virtually all types of non-marine vegetation. Indeed if we count the *genets* above and below 1 m^2 of almost any soil surface, many more individuals lie dormant below the surface than grow above it. Some typical seed population sizes for various types of

Fig. 2.2 Frequency distributions of dispersed seeds of four arctic alpine herbs. (From Marchand and Roach 1980)

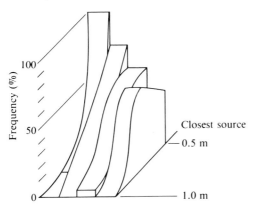

Table 2.2 Numbers of seeds and the predominant species present in the seed pools of various vegetation types

Vegetation type	Location	Seeds m^{-2}	Predominant species in the soil	Source
Tilled agricultural soils				
Arable fields	England	28 700–34 100	Weeds	Brenchley and Warrington 1933
Arable fields	Canada	5000–23 000	Weeds	Budd, Chepil and Doughty 1954
Arable fields	Minnesota, USA	1000–40 000	Weeds	Robinson 1949
Arable fields	Honduras	7620	Weeds	Kellman 1974b
Grassland, heath and marsh				
Freshwater marsh	N. Jersey, USA	6405–32 000	Annuals and perennials representative of the surface vegetation	Leck and Graveline 1979
Salt marsh	Wales	31–566	Sea rush where abundant in vegetation grasses	Milton 1933
Calluna heath	Wales	38 000	*Calluna vulgaris*	Chippendale and Milton 1934
Perennial hay meadow	Wales	18 875–19 625	Subsidiary species of the vegetation	Chippendale and Milton 1934
Meadow steppe (perennial)	USSR	2000–17 000	Annuals and species of the vegetation	Golubeva 1962
Perennial pasture	England	300–800	Subsidiary species of the vegetation, many annuals	Champness and Morris 1948
Prairie grassland	Kansas, USA	1980	*Zoysia japonica*	Lippert and Hopkins 1950
Zoysia grassland	Japan	18 780	*Miscanthus sinensis*	Hayashi and Numata 1971
Miscanthus grassland	Japan	9000–54 000	Annual grasses	Hayashi and Numata 1971
Annual grassland	California, USA			Major and Pyott 1966
Forests				
Picea abies (100 yrs old)	USSR	1200–5000	All earlier successional spp.	Karpov 1960
Secondary forest	N. Carolina, USA	1200–13 200	Arable weeds and spp. of early succession	Oosting and Humphries 1940
Primary subalpine conifer forest	Colorado, USA	3–53	Herbs	Whipple 1978
Subarctic pine/birch forest	Canada	0	No viable seeds present	Johnson 1975
Coniferous forest	Canada	1000	Alder *Alnus rubra*	Kellman 1970
Primary conifer forest	Canada	206	Shrubs and herbs	Kellman 1974a

vegetation are given in Table 2.2. Though these seed populations are of some interest in themselves, ultimately they are only of importance from the point of view of the plant when seeds are recruited from the seed pool to the growing population. In the light of this it is interesting to compare the species composition of seed pools with that of the active population.

In the majority of cases when buried seeds are identified and counted it is found that there is little direct relation between the abundance of a species above ground and the abundance of its seeds in the soil (Table 2.2 and Fig. 2.3(a)). Forest soils in late successional woodland predominantly contain the seeds of plants of earlier stages in the succession, and perennial grassland soils are replete with the seeds of short-lived species and contain few seeds of the dominant grass species. The exceptions to these discrepancies between the growing flora and the species occurring in the seed pool occur in arable fields (Fig. 2.3(b)) and in the seed pools of annual grasslands such as those studied by Major and Pyott (1966) in California.

These exceptions suggest an explanation for the paradoxical composition of most seed pools. It is that dormant seeds are only produced in large numbers by species whose growing populations are subject to periodical local extinction. This is plainly the case for early successional plants, grassland annuals and arable weeds, all of which are usually well represented in the soil of their respective habitats.

Though the small seeds of short-lived plants dominate seed pools beneath many types of vegetation it is incorrect to assume that all short-lived plants rely upon buried dormant seed for recruitment. *Vulpia fasiculata* is a small annual grass of coastal sand dunes which possesses no seed dormancy. Seeds of this species germinate soon after they ripen and very few apparently become incorporated into a seed pool (Watkinson 1978a). Sterile brome grass (*Bromus sterilis*) is another annual of waste places in Britain with virtually no seed dormancy (Chancellor 1968). The lack of seed dormancy in *Vulpia* may explain why it is confined to the special environment of shingle and sand-dune habitats (Watkinson 1978b) and why it is not found in arable habitats, but *B. sterilis* has recently become a serious arable weed.

Despite the fact that the species represented most abundantly in seed pools often rely heavily or even entirely on this source of seeds for recruitment, the limited evidence available suggests that only a very tiny fraction of these seeds ever produce seedlings. Roberts and Ricketts (1979) found that the total weed seedling numbers present in fields after cultivation represented only 3–6 per cent of the weed seeds present in the top 10 cm of the soil before the weeds emerged. Naylor (1972) estimated the fraction of emerging seedlings of a weed grass *Alopecurus myosuroides* which were derived from seeds which had been in the soil longer than 1 year by a mark and recapture experiment. Seeds of the *A.*

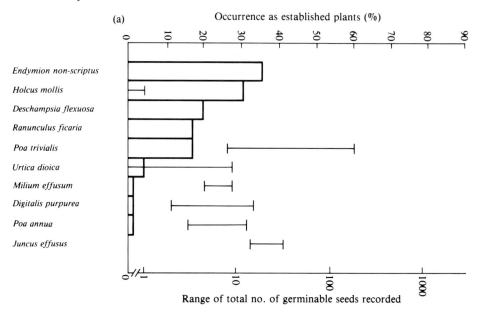

(a)

Occurrence as established plants (%)

Endymion non-scriptus

Holcus mollis

Deschampsia flexuosa

Ranunculus ficaria

Poa trivialis

Urtica dioica

Milium effusum

Digitalis purpurea

Poa annua

Juncus effusus

Range of total no. of germinable seeds recorded

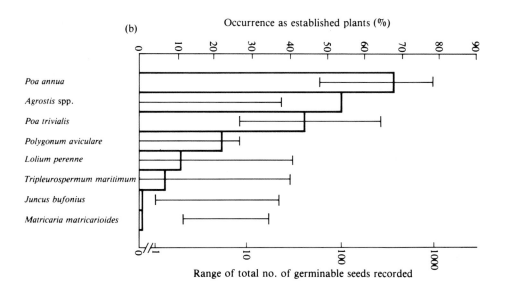

(b)

Occurrence as established plants (%)

Poa annua

Agrostis spp.

Poa trivialis

Polygonum aviculare

Lolium perenne

Tripleurospermum maritimum

Juncus bufonius

Matricaria matricarioides

Range of total no. of germinable seeds recorded

Fig. 2.3 The relative abundance of mature plants (histograms) and of seeds in the soil (vertical bars) for the major species of (a) a deciduous woodland and (b) an arable field. (From Thompson and Grime 1979)

myosuroides were marked with a dilute solution of fluorescent paint
and the seed production of plants in field plots was estimated. Quantities of marked seeds equal to one-tenth the number of naturally
dispersed, unmarked seeds were added to the plots before the fields
were harvested. When seedlings of *A. myosuroides* emerged in the
experimental plots in the following season they were excavated to reveal
the remains of the attached seed. The number of seedlings with a
marked seed attached was equal to one-tenth of all the seedlings
produced from seeds dispersed the previous year in the plot. Naylor
calculated from this that 60–70 per cent of emerging seedlings were
derived from the previous season's seed production. This figure suggests
that, at least in this weed species, the seed pool does not provide many
recruits to the growing population from seed that has been dormant for
a long time.

There is an important difference between the cohort as defined in
studies of animal populations and the way this term is used in practice
for plant populations recruited from seed. Once an animal is born it
generally embarks upon juvenile development straight away. Although
there are some exceptions to this such as the extended diapause which
may be found in insects, most metazoan populations possess no equivalent of the seed pool which may introduce a delay of indeterminate
duration between the birth of a seed and its emergence as a seedling.
Thus, unlike an animal cohort, a cohort of emerging seedlings may
consist of a collection of individuals which were not actually born at the
same time.

This may be important if there are several different genotypes among
seeds and if natural selection is acting upon the population. In these
circumstances different parental genotypes may contribute different
quantities of seed to the seed pool in different years. When each new
cohort of seedlings is drawn evenly from this pool of mixed origin it will
contain genotypes derived from several different seasons and not just
those most favoured in the most recent season. This effect of the seed
pool may *buffer* genetic changes in plant populations. Epling, Lewis and
Ball (1960) observed that this effect buffered year-to-year changes in the
frequency of different flower colours in *Linanthus parryae*, an annual
which occurs in the Mojave desert.

The fates of buried seeds

The fate of seeds in the soil is difficult to determine and comprehensive
information on the dynamics of particular species' seed pools is sparse.
In an experimental approach to the problem, Sarukhan (1974) sowed
replicated samples of 100 viable seeds of three buttercups, *Ranunculus
repens*, *R. bulbosus* and *R. acris*, into small areas of grassland in which

natural seed dispersal was prevented. The fate of these seeds was classified in the five categories illustrated in Fig. 2.4 by counting emerging seedlings and retrieving buried seeds at intervals through the year. Seeds recovered from the soil were subject to a germination test to detect dormancy and the ungerminable fraction was then tested for viability using tetrazolium chloride which stains living tissue red (Smith and Thorneberry 1951). The fates of seeds of *R. repens* and *R. bulbosus* determined by these methods over a 15-month period is shown in Fig. 2.5. Rodents were responsible for the heavy predation experienced particularly by *R. repens* which lost 50 per cent of its seed pool in the first 6 months of the experiment. Other deaths occurred from unknown causes. Even though *R. repens* lost more seeds to predators than *R. bulbosus*, the seed pool of the latter species was depleted more rapidly as a result of substantial germination.

Seed predation by rodents is also an important factor in depleting the seed pool of the annual grasses which form some of the commonest species in Californian annual grassland. Borchert and Jain (1978) estimated the number of seeds taken by wild house mice (*Mus musculus*) and California voles (*Microtus californicus*) by excluding these predators from plots sown with known quantities of seeds of wild oats (*Avena fatua*), wild barley (*Hordeum leporinum*), ripgut brome (*Bromus diandrus*) and Italian ryegrass (*Lolium multiflorum*). *Avena fatua* seeds were preferred over those of the other species and 75 per cent of their seeds were eaten. Mice and voles consumed 44 per cent of the seeds of *H. leporinum* and 37 per cent of *B. diandrus*. The ultimate size of the population of growing plants in all of these species was not affected as severely as these levels of predation on seeds would suggest because a partly compensatory reduction in density-dependent mortality occurred

Fig. 2.4 A model of the dynamics of the seed pool. (After Harper 1977)

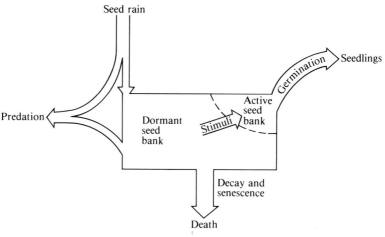

Fig. 2.5 The fates of seed samples of (a) *Ranunculus acris*, (b) *R. repens* and (b) *R. bulbosus*. (From Sarukhan 1974)

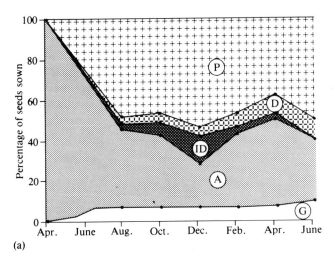

(a)

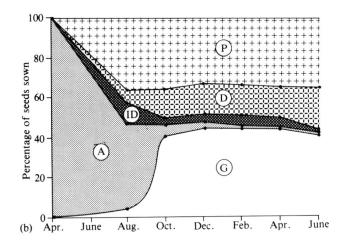

(b)

(P) Predation

(D) Decay

(ID) Dormant

(A) Active

(G) Germination

in the plant populations from which seeds were removed by rodents. The survival from seeds to adult plants of *Avena, Hordeum* and *Bromus* was greater in the presence of seed predation than in its absence inside exclosure fences.

Granivorous rodents appear to be particularly important in depleting the seed pools of desert shrubs and herbs (Brown, Reichman and Davidson, 1979). An estimated 30–80 per cent of the seed losses observed by Nelson and Chew (1977) in a study in the Mojave desert were attributed to rodents, the most important species being the pocket mouse (*Perognathus formosus*). Nelson and Chew estimated that only 25 per cent of seed losses were accounted for by germination, and they concluded that the total losses from the seed pool were not large enough to prevent a net accumulation of seeds in the soil. Harvester ants are also major predators of seeds in North American deserts (Brown, Reichman and Davidson 1979).

Ants and other seed predators may operate as dispersers of those seeds which they carry away but do not consume. Both rodents and birds cache seeds during periods of abundance. Not all of these may be recovered before they germinate. The pinyon pine (*Pinus edulis*) in North America and oak (*Quercus robur*) in the Netherlands both depend on regeneration from seeds cached by jays (Ligon 1978; Bossema 1979).

Seed dispersal by ants is more than an incidental act in the course of predation for certain plant species such as violets (*Viola* spp.). These have a special oil-containing appendage attached to the seed, known as an elaiosome, to which ants are attracted. Seeds with an elaiosome are carried off to the nest, the elaiosome is eaten and the seed itself, still viable, is deposited on the ants' refuse heap where it may germinate. In a study of ant/seed interactions (known as myrmecochory) in a group of violets growing in a forest in West Virginia, Culver and Beattie (1978) found that one species (*Viola papilionacea*) depended on ant transportation of its seeds to escape bird and rodent predators which ate any seeds not removed by ants. Harvester ants appear to afford similar protection from rodents to *Datura discolor* in the Sonoran desert (O'Dowd and Hay 1980).

Experiments on the decay of viable weed seed populations in cultivated soil by Roberts and Feast (1973) have shown that seed survivorship in the soil declines exponentially. However, seed survival is higher in uncultivated soil and these experiments may not reflect the situation for plants of other habitats. Odum (1978) selected 100 sites of abandoned settlements, abandoned farmland and demolished buildings in Denmark which were dominated by perennial vegetation and exposed the underlying soil to allow buried seeds to germinate. At virtually every site large numbers of annuals, biennials and other short-lived plants grew from the soil. The most spectacular example of seed survival in this study occurred in the soil removed from the excavation of an eleventh-century grave from which a plant of *Verbascum thapsiforme* germinated after 850 years of dormancy.

Seed dormancy

An enormous number of studies of seed dormancy have been conducted, mostly by physiologists attempting to elicit the mechanisms which inhibit and trigger germination (Mayer and Polijakof-Mayber 1975). These studies have revealed a bewildering variety of factors influencing germination in different species including light intensity, photoperiod, light quality (spectral composition), temperature, temperature fluctuations, nitrates, O_2 and CO_2 levels, pH, moisture and physical abrasion of the seed coat, to name only the most common ones. To complicate matters still further, there is evidence that the conditions under which seeds are stored can induce dormancy in seeds which show no dormancy when freshly collected. There are also several examples of polymorphism within populations and geographic variation within species for seed dormancy. These show that many physiologically oriented studies carried out on seeds of unknown provenance, are of limited value to the population ecologist who would like to draw general conclusions from them. Most of the earlier studies of seed germination which are of interest to ecologists were carried out by weed researchers (Roberts 1970).

Despite these reservations about the evidence concerning seed dormancy in particular species, it is possible to sum up the general situation quite simply: 'Some seeds are born dormant, some achieve dormancy and some have dormancy thrust upon them' (Harper 1959). These three types of dormancy are termed *innate*, *induced* and *enforced* dormancy respectively (Harper 1977), and play slightly different roles in the regulation of germination. Fresh, innately dormant seed will not germinate in the conditions of normal germination tests (warmth and moisture on a filter paper or agar substrate) and will lie dormant in the seed pool till they receive a specific stimulus to break dormancy. Some period of 'after-ripening' may be required before the stimulus will have effect. Umbellifer seeds possess innate dormancy which is broken by a period of exposure to cold (stratification) when they are in the imbibed (water-saturated) state. Umbellifers such as cow parsley (*Anthriscus sylvestris*), hogweed (*Heracleum sphondylium*), wild angelica (*Angelica sylvestris*), pignut (*Conopodium majus*) and wild parsnip (*Pastinaca sativa*) which occur in Britain germinate in the spring after they have received the appropriate stratification and after the soil has warmed up sufficiently (Roberts 1979).

Dormancy may be induced in some seeds which are born without it by burial in the soil or by exposure to light filtered through a canopy of leaves. Wesson and Wareing (1969a, b) demonstrated that seeds of several species (*Chenopodium rubrum, Plantago lanceolata, Polygonum persicaria, Spergula arvensis* and others), which show no dormancy or germination requirement for light when freshly collected, occur in a dormant state in the seed pool and that this dormancy is broken by light when the soil is disturbed. A number of grassland species including several perennials produce seeds with little or no sign of innate dormancy when germinated in the dark but show induced dormancy when placed under a canopy of leaves (Silvertown 1980a) in response to the increased ratio of red/far-red light which occurs beneath leaves. These types of induced dormancy ensure that seeds do not germinate in circumstances where they will encounter a barrier of soil or vegetation which would reduce the survival of seedlings. In contrast to innate dormancy which locates seed germination in time, dormancy induced by burial or a leaf canopy locates or restricts seed germination in space as well as time.

Seeds in enforced dormancy may be released from this state simply by providing adequate water to allow imbibition at a normal temperature. Seeds held in enforced dormancy for a sufficient length of time may develop induced dormancy if they become buried, and combinations of innate, induced and enforced dormancy are also commonly found. Fresh seeds of knotgrass (*Polygonum aviculare*) show innate dormancy which prevents germination before the winter but this is broken by stratification and seedlings emerge in the spring. Seeds which have broken their innate dormancy but remain in enforced dormancy till the end of May then acquire renewed (induced) dormancy which prevents germination until another period of winter cold has passed (Courtney 1968; Roberts 1970). A similar seasonal adjustment of dormancy is achieved by a different means in fat hen (*Chenopodium album*) which produces non-dormant seeds early in the season and innately dormant ones in the main crop.

Many species show phenotypic polymorphism for seed dormancy and produce both dormant and non-dormant seeds in the same seed crop. Arthur, Gale and Lawrence (1973) studied cohorts of seedlings produced from both dormant and non-dormant seeds of *Papaver dubium*, a poppy which is an arable weed in British fields. Seedlings emerging in the autumn from non-dormant seeds were prone to heavy mortality over the winter in some years, but the survivors of autumn cohorts produced larger plants and at least ten times more seeds than seedlings emerging in the spring.

Many annual species have a flush of seedlings in more than one season of the year. It may well be common in these populations for seed production to be relatively greater in the earliest cohort and seedling survival relatively greater in the later one. This pattern of seed production and seedling survivorship has been observed in autumn- and spring-germinating cohorts of prickly lettuce (*Lactuca serriola*) (Marks and Prince 1981) in Britain, and in summer and autumn cohorts of *Leavenworthia stylosa* in Tennessee cedar glades in the USA (Baskin and Baskin 1972). Autumn cohorts of *Lactuca* experience 1.3 times the mortality of spring ones because of winter temperatures. Summer cohorts of *Leavenworthia* experience over four times the mortality of autumn ones because of drought. Relative seed production is in favour of the earlier cohort in both species. Survivors of the autumn cohort of *Lactuca* produced twice the number of seeds produced by survivors of the spring cohort. Survivors of the summer cohort of *Leavenworthia* produce nearly eight times more seed per surviving plant than autumn cohorts.

These populations illustrate the role seed dormancy may play in increasing seedling survival by delaying germination to a later season. They also demonstrate that this delay may carry a penalty in terms of reduced seed production when compared with the seed production of survivors of an earlier season's cohort. In species with a widespread geographical distribution the seasonal hazards of germination are likely to vary with latitude. Red campion (*Silene dioica*) populations of southern Europe produce seeds with dormancy characteristics which allow them to avoid summer drought, and seed populations from northern Europe are winter-dormant (Thompson 1975).

When the survivorship of seedlings emerging early and late within the same season's cohort is compared (Fig. 2.6), the first seedlings to emerge are generally at an advantage. The effects of even a slight delay in seedling emergence may be far-reaching for the subsequent fate of the plant. In a wood where he was studying the sweet white violet *Viola*

Fig. 2.6 The percentage of seedlings surviving from cohorts emerging at successive intervals in natural populations of (A) *Androsace septentrionalis* and (B) *Tragopogon heterospermus*. (From Symonides 1977)

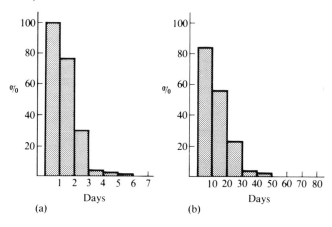

blanda, Cook (1980) found a cohort of newly emerged seedlings in among a cohort he had marked 15 days before. He marked these also and then noted that the average size of seedlings from the late cohort remained consistently smaller than for the earlier ones during the following 3 years of their life. Smaller plants experienced a greater overall risk of mortality than larger ones. During a period of high mortality which occurred in the third year, plants of the later cohort suffered significantly greater mortality than plants of the earlier one which were less than 15 days their senior in age.

Trends of this kind could either be due to environmental conditions deteriorating so that seedlings emerging later are adversely affected by the weather, or it may be due to older, larger seedlings capturing a larger share of resources and suppressing new ones as they emerge. Weaver and Cavers (1979) characterized these two alternatives as the effects of emergence *order* (e.g.: whether seedlings are the first or last to

Fig. 2.7 The percentage contribution of successive (1st, 2nd, 3rd) cohorts of seedlings of *Rumex crispus* to the final number of plants in three populations sown at different times. (Redrawn from Weaver and Cavers 1979)

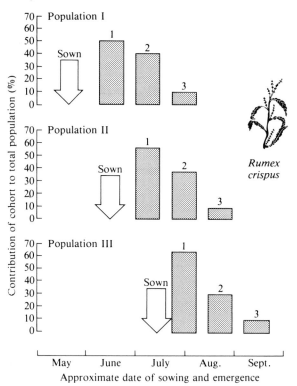

emerge) and the effects of emergence *date* (e.g.: March or April emergence) on seedling survival. They tested the relative importance of these two factors in populations of curl-leaved dock (*Rumex crispus*) and broad-leaved dock (*R. obtusifolius*) by sowing batches of seed at monthly intervals and comparing the relative contribution to the final dock population made by seedlings emerging in the first, second and third intervals in each period of seed germination. Percentage mortality was highest in the cohorts of later emergence order and their results (Fig. 2.7) showed that the order of emergence was more important than the date of emergence. This suggests that the accurate timing of seed germination and the rapid breaking of seed dormancy is essential to minimize the effects of interference from other seedlings.

The safe site

A knowledge of the germination responses of seeds in the laboratory is not sufficient to predict accurately when and where a seed is capable of germination in the field. Simple ecological methods used in a number of studies have demonstrated that seed germination is highly responsive to fine-scale differences in the physical environment at the soil surface and that physiological studies of dormancy tell only half the story of why seeds do or do not germinate in the field. In a study of the response of seeds to microtopographical variation in the soil surface, Harper, Williams and Sagar (1965) sowed a seed bed with equal proportions of the seeds of three plantains *Plantago media*, *P. lanceolata*, and *P. major*. The seed bed was divided into sub-plots and various objects were then placed on the surface of these. Treatments were replicated in a randomized design. The treatments, listed in the legend of Fig. 2.8, had selective effects on the seedling emergence of the different species which showed marked differences in their response to different conditions of soil microtopography and micro-environment.

Harper, Williams and Sagar (1965) adopted the term *safe site* to describe those specific conditions in the soil surface which permit seeds to escape all the hazards of the pre-germination phase (including predation) and to overcome dormancy. Although this term shares with the term *niche* the unfortunate property that a particular species' safe site can only be identified *after* it has been successfully occupied, the term provides a useful conceptual handle with which to grapple with the idea that a multitude of factors affect seed germination.

The shape and size of a seed in relation to the soil particles of the surface in which it rests seem to play an important part in determining the availability of the water needed for successful germination. Oomes and Elberse (1976) compared the seed germination of six grassland species when sown on an even soil surface and in 10 mm and 20 mm wide grooves in the surface of the soil. Yarrow (*Achillea millefolium*)

Fig. 2.8 (a) The distribution of various treatments to the *Plantago* seed bed: (1) and (2) two kinds of depression in the soil surface; (3) a sheet of glass laid on the soil surface; (4a) and (4b) sheets of glass placed vertically in the soil; (5), (6) and (7) rectangular wooden frames of three different depths pressed into the soil surface; x worm casts. (b) the distribution of seedlings of *P. lanceolata*, (c) *P. media* and (d) *P. major*. (From Harper, Williams and Sagar 1965)

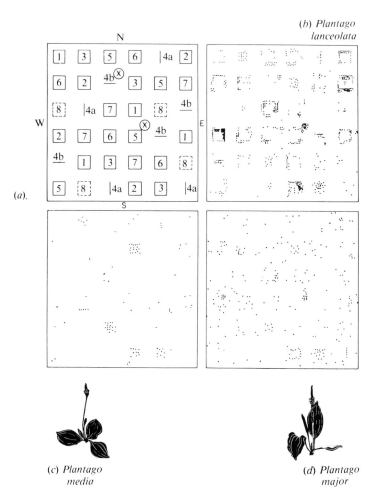

(*b*) *Plantago lanceolata*

(*a*).

(*c*) *Plantago media*

(*d*) *Plantago major*

has flat seeds which germinated best on the even surface, but several
species with other shapes (e.g. ox-eye daisy *Chrysanthemum leucanthe-
mum* and self-heal *Prunella vulgaris*) germinated poorly when lying in
this position and did much better in the 20 mm grooves. Perhaps one of
the most important factors which determines the spatial pattern of
seedling emergence and recruitment in the field is the microdistribution
of the leaves and leaf canopies of established plants. Oak seedlings of
Quercus robur are light demanding and are also prone to defoliation by
moth caterpillars which can descend on them from established trees.
Either or both of these factors may explain why oak saplings are not
often found beneath the canopy of large oaks (Fig. 2.9). Interactions
with a similar result occur among herbaceous plants. The distribution of
seedlings of a short-lived perennial *Reseda lutea* in relation to the leaf
canopy in a grassland habitat is shown in Fig. 2.10. The explanation for
this particular pattern is not known.

Fig. 2.9 The distribution of young oaks around a mature tree at Silwood
Park in Berkshire. (From Mellanby 1968)

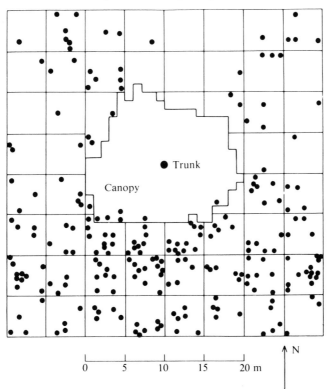

Summary

A *life table* lists the age-specific probabilities of mortality and survival for a cohort of individuals. A *fecundity schedule* lists age-specific reproductive rates in parallel to the life table. *Dynamic* life tables are drawn up by following the fate of a cohort. *Static* life tables are derived by more indirect methods.

Because of the plastic growth of plants, individuals of the same *chronological age* may be at quite different *stages* of growth. Particularly in monocarpic plants, flowering is often more closely tied to an individual's growth stage than to its age.

Dormancy is found in a variety of growth stages during plant life cycles. Seedlings (oskars), tubers and rhizomes are variously employed to postpone development between seed germination and flowering. Dispersal is a possible alternative to dormancy as a response to changes in the environment. Seeds are rarely dispersed far from the parent in large numbers, although the few seeds which do travel a long way may become the founders of new populations.

Seeds are the commonest form of dormancy organ and are found in enormous numbers beneath many vegetation types. There is generally a discrepancy between the species found above ground and those represented most abundantly in the soil. These are predominantly species with a short lifespan and small seeds. Because the seed pool accumulates seeds from many successive generations it may potentially buffer genetic changes in populations which regularly recruit from seed.

Fig. 2.10 The distribution of seedlings of *Reseda lutea* in relation to the microdistribution of vegetation in a 25 cm square plot of chalk grassland. The density of hatching is proportional to percentage vegetation cover. (From Silvertown 1981a)

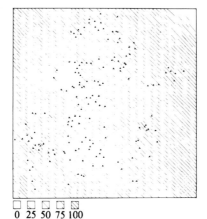

0 25 50 75 100
% vegetation cover

Very few buried seeds ever produce mature plants. Many are eaten by vertebrate and invertebrate predators. Some predators may also act as seed dispersers during the process of caching or transporting their food. Seed numbers in undisturbed soil decline exponentially. This decline is speeded up by soil disturbance.

Seed dormancy may be broken by a wide variety of factors in different species. Three general types of dormancy may be distinguished: *innate*, *induced* and *enforced*. Some of these may occur alone, successively in time or in the same seed and in different seeds in the same population. The breaking of dormancy and the timing of germination have measurable effects upon the subsequent survival and reproduction of seedlings.

Various physical environmental factors determine whether a seed will germinate in the soil and how the surviving seedlings will be distributed. The term *safe site* has been given to those conditions which permit successful establishment of a particular species from seed.

3
The demography of some plant populations

Demography is the corner-stone of a variety of studies designed to serve many different ends. The fundamental nature of the information contained in the life table and fecundity schedule means that it is of interest from many points of view, both theoretical and applied.

The market gardener is interested in how a cabbage crop succumbs to mortality through the season and interprets this information in terms of the loss of market value (Fig. 3.1). A forester's interest in the survivorship and mortality of trees is similarly motivated (see Fig. 3.20, p. 68). Conservationists and foresters practising some kinds of management are both interested in monitoring (and perhaps increasing) the rate of natural replacement which occurs in plant populations. Agronomists may require similar information to manage pastures and rangelands. Both may also be interested in the size of the fraction of a self-renewing population which may be harvested without harming it. Wildlife man-

Fig. 3.1 The survivorship of a commercial cabbage crop. The main causes of mortality and the cost of this in lost market value are shown. (Drawn from data of Harcourt 1970)

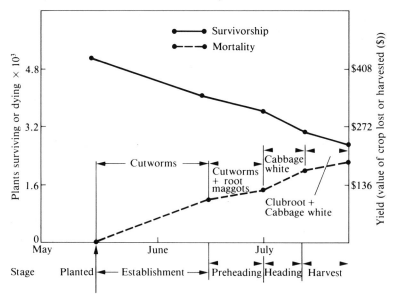

agers have spent considerable effort collecting information on the fruit-ing habits of wild shrubs and trees which provide food for animals.

Population ecologists interested in more theoretical questions about plant populations depend heavily upon the results of these applied studies. One of the functions of the population ecologist is to produce useful generalizations. Demography may help to do this in two areas. Firstly, it supplies information on the dynamics of populations, includ-ing the rate at which one individual is replaced by another. Secondly, where data are available on genetic differences between individuals, it may supply information on the relative replacement rate of different genotypes. Hence demography is also the corner-stone of studies in population genetics and evolution.

Two methods exist for obtaining the information required to fill out any life table with values of age-specific survivorship (l_x) and age-specific mortality (d_x). The first is simply to follow the demise of a cohort of seedlings, taking censuses at intervals. This is the only method available for analysing populations with non-overlapping generations (or cohorts) such as those of *Phlox drummondii* discussed earlier (p. 15). A life table produced by this method is known as a *dynamic* or *horizontal* life table. It is often impractical to make a census of cohorts of long-lived organisms such as trees throughout their lifetime and a short-cut method must be used. In these cases values of l_x and d_x are calculated from the *age structure* of the population at a single (rarely at two) sampling date. Life tables produced by this method are known as *static* or *vertical* life tables. Assumptions have to be made about recruitment to populations analysed in this way and static life tables are consequently more prone to error. Their use is discussed separately in this chapter on p. 58.

The patterns of mortality observed in different plant populations appear to vary as a result of two factors: firstly, species appear to have roughly characteristic patterns of mortality which depend upon whether they are annuals or perennials, herbs or trees. This is hardly surprising since these colloquial terms describe life history patterns which must be reflected to some extent in the life table. Secondly, their mortality patterns are modified by the particular circumstances prevailing in the different habitats which populations of the same species occupy. The density of a population is one important factor which affects mortality (Ch. 5).

Survivorship curves can be crudely classified into three types depend-ing on the distribution of mortality risk with age. These three types of curve, illustrated in Fig. 3.2, are often named Deevey types I, II and III after one of the first authors to compare life tables for different animal species (Deevey 1947). A cohort with type I survivorship has low mortality in early and middle life but a rapid change to high mortality later on. Type II survivorship is typified by a constant death risk throughout, and type III is a pattern of high juvenile and low adult mortality.

A useful statistic which describes some characteristics of survivorship is the half-life of a cohort. Half-life is an appropriate measure of the rate of mortality only in Deevey type II populations, or for those sections of some other type of curve which show a constant death risk, and hence an exponential decline in numbers. It may be calculated from the formula:

$$\text{Half-life in years} = \frac{t \ln 2}{\ln N_x - \ln N_{x+t}}$$

where N_x is the number of survivors at age x and N_{x+t} is the number remaining after a time interval of t years.

The demography of monocarpic herbs

Several studies of survivorship in annual species have described mortality between germination and seed set conforming remarkably well to Deevey's type I curve.

Virtually all of the annual populations whose seedling survivorship curves are depicted in Fig. 3.3 grow in habitats which are subject to annual or more frequent disturbance. *Alyssum* and *Leavenworthia* colonize bare soil, *Cerastium* is a plant of sand dunes and *Salicornia* occurs on tidal mud. The obvious hazards of these habitats do not appear to contribute significantly to the major episode of mortality in these populations which occurs after flowering. Two exceptions are *Sedum smallii* and *Minuarta uniflora*, both annuals growing on rock outcrops where seedlings are liable to be washed away by rain.

It is unwise to attempt to generalize about survivorship for particular

Fig. 3.2 Model survivorship curves. Deevey types I, II and III.

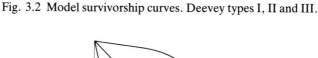
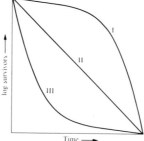

species from only 1 year's data, but some tentative generalizations may be reached by comparing studies of different species. Annuals tend to occupy ephemeral sites which, though open to colonization in one year, may be closed to it in the next. It seems likely that the significant hazards affecting annual populations are not perpetual disturbance,

Fig. 3.3 Some representative survivorship curves for annual plants:
(1) *Alyssum alyssoides* (Baskin and Baskin 1974); (2) *Cerastium atrovirens* (Mack 1976); (3) *Leavenworthia stylosa* (Baskin and Baskin 1972);
(4) *Phlox drummondii* (Leverich and Levin 1979); (5) *Sedum smalii* and (7) *Minuarta uniflora* (Sharitz and McCormick 1975); (6) *Salicornia europaea* (Jeffries, Davy and Rudmik 1981).

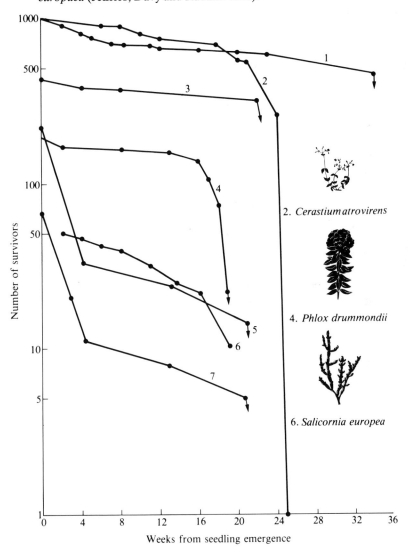

2. *Cerastium atrovirens*

4. *Phlox drummondii*

6. *Salicornia europea*

every day of the week, but the uncertainty of what next year will bring. In other words, the probability of seeds broadcast randomly landing at a suitable site of recent disturbance is low and unpredictable, but once such sites have been reached seedling mortality in that growing season at those sites is minimal. Note that the survivorship curves in Fig. 3.3 do not extend over the seed stage of the life cycle. Mortality during this stage accounted for 80 per cent of seeds produced in *Phlox* and *Sedum* and 94 per cent in *Minuarta*.

Because they colonize sites of local disturbance, annuals often occur in dense stands. This sometimes produces density-dependent seedling mortality. Annuals, like other plants, may also react to crowded conditions by plastic reductions in plant size which may produce a density-dependent adjustment of fecundity (Ch. 5). This was observed in *Salicornia, Sedum* and *Minuarta* populations and may have occurred in the other species too, although this was not investigated.

Survivorship curves for three monocarpic perennial herbs are shown in Fig. 3.4. These plants are also colonizers of disturbed sites but delay seed production until the second or even later years of growth. Herbs behaving in this way are often described by the misnomer *biennial*. *Melilotus alba* is a true biennial, and all individuals in the population studied by Klemow and Raynal (1981) flowered or died by the second year. Individual plants of *Grindelia* and *Pastinaca* may delay flowering beyond their second season. Flowering in these species depends upon rosettes reaching a minimum size (see Ch. 2).

The delayed reproduction found in monocarpic perennials places constraints on growth and seed production which do not exist for annuals which complete their life cycle within 12 months. Whereas an annual may succeed in colonizing a gap and reproducing in it before it is closed by the invasion of surrounding perennial vegetation, perennial monocarps must contend with the interference of invading plants during two seasons of growth at least. Since most gaps begin to close very soon after they are created, the time a biennial arrives in such a gap is critical to its final success in producing seeds. This has been demonstrated for mullein (*Verbascum thapsus*) (Gross 1980) and wild carrot (*Daucus carota*) (Holt 1972) colonizing old fields in the USA. Both species may germinate in patches of bare soil where vegetation cover is reduced, but the probability of a rosette finally reaching sufficient age or size to flower is dependent upon how much interference is present from other vegetation.

By the time a monocarpic perennial does flower, the site which saw its birth has usually become closed to successful establishment by that plant's progeny. Harper (1977) has remarked that when biennials at a particular site attract the attention of an ecologist by their flowering, they are generally about to become locally extinct. Looking at this from the plants' point of view, one could say that the appearance of an ecologist in the area is an ill omen! *Chaerophyllum prescottii*, the monocarpic perennial mentioned in Chapter 2 p. 21, has overcome the problem of re-establishment by synchronizing seed production with events which generate new gaps. *Verbascum thapsus*, foxglove (*Digitalis purpurea*) and many other monocarpic perennials have large seed pools and employ persistent seeds as a means of colonizing gaps as the opportunity arises.

Many monocarpic umbellifers have neither a seed pool nor the

Fig. 3.4 Survivorship curves of three monocarpic perennials:
(1) *Grindelia lanceolata* (Baskin and Baskin 1979a); (2) *Pastinaca sativa* (Baskin and Baskin 1979b); (3) *Melilotus alba* (Klemow and Raynal 1981)

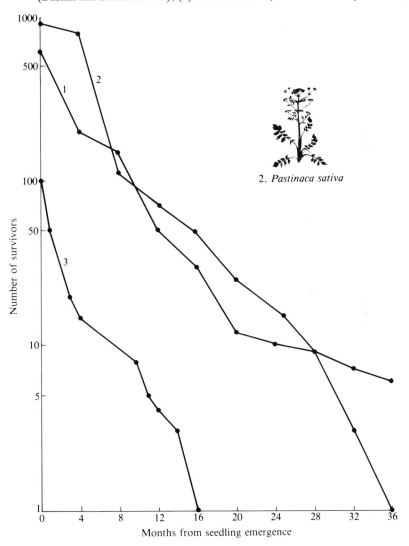

2. *Pastinaca sativa*

Number of survivors

Months from seedling emergence

persistent tuber of prescott chervil and are consequently confined to habitats in which regular disturbance permits regular re-establishment from the previous year's seed. Siberian hogweed (*Heracleum sibiricum*) is among several species of flood meadows near the river Oka in the USSR which depend upon this source of recruitment (Rabotnov 1978). The life cycle of these plants is something like a relay race. No period of inactivity in the soil is permitted between one generation and the next. The occasions when whole populations become extinct because disturbance fails to occur after flowering must be reduced to some extent by the perennial habit and the variable time plants of the same cohort may take to reach the flowering stage.

Observations of survivorship in both monocarpic and polycarpic perennials never seem to show type I curves, although both types II and III are found. In a study of the demography of eleven sand dune species in the Torun basin in Poland, Symonides (1977) found juvenile mortality to be higher in two 'biennial' species (80%) than among five annuals (60–70%). This follows the trend suggested by our other examples. Four perennial, polycarpic species in Symonides' study experienced between 85 and 97 per cent seedling mortality before flowering.

The demography of polycarpic perennial herbs

The populations in Symonides' study occupied a successional environment. During the 8 years of her observation, several of Symonides' sites showed trends in population size for individual species which are probably typical of processes operating in changing herbaceous plant communities. One of the pioneer grasses which fixes mobile sand is grey hair grass (*Corynephorus canescens*) (Symonides 1979a).

At three sites where this plant was studied, seedlings were found in abundance each year, clumped around established individuals (tussocks) of the species. The seed distribution of this species in the soil was also clumped around tussocks (Symonides 1978). Figure 3.5(a) shows the distribution of seedlings and tussocks at a site newly colonized in 1968, the first year of the study. Seedling mortality was high during the year (Fig. 3.5(c)), particularly among seedlings emerging late relative to the rest of the cohort and those some distance from established tussocks.

Despite this mortality, some recruitment to the population of established plants did occur. Figure 3.5(b) shows how large annual flushes of seedlings produced annual increments in the established population,

Fig. 3.5 The population dynamics of *Corynephorus canescens*. (Data from Symonides 1979a)

(a) Map of tussocks and seedlings. (b) Seasonal fluctuations in total population density and annual density of adult plants. (c) Annual percentage mortality rate of seedling and non-seedling individuals. (d) Percentage of fruiting individuals in the non-seedling

population annually. (e) Annual fecundity of the adult population
measured as seedlings/tussocks and seeds/tussocks. (f) Age structure of
established plants.

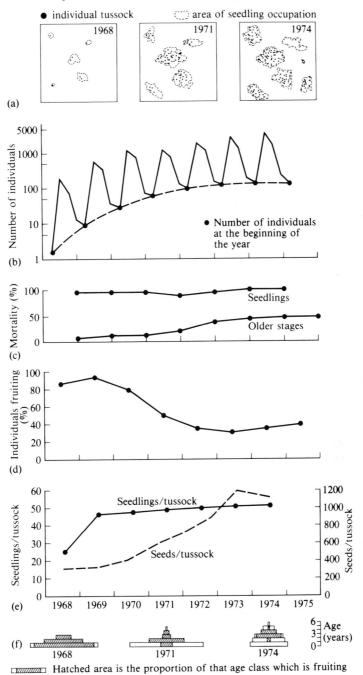

(a)

(b)

(c)

(d)

(e)

(f)

Hatched area is the proportion of that age class which is fruiting

which eventually stabilized at a density of 170 tussocks in the 4 m² plot. The sequence of diagrams in Fig. 3.5(a) shows that recruitment in this population leads to gradual accretion of new individuals around the oldest tussocks.

As the process of accretion progressed, the age structure of established tussocks developed from a juvenile to a senescent form (Fig. 3.5(f)). This change in age structure occurred remarkably quickly compared to the average lifespan of an established plant which Symonides calculated to be about 7.8 years at this site. At two other sites where the colonization process had begun earlier and hence progressed further, senescence was further advanced, with a greater proportion of individuals in older age classes.

In the early stages of population growth, a high proportion of established individuals fruited (Fig. 3.5(d)). As population size increased, tussocks in younger age classes fruited less frequently than older ones because of the effects of crowding. By 1975 reproduction was practically confined to tussocks more than 2 years old (Fig. 3.5(f)). However, seed production per tussock increased with time. The number of seedlings produced per tussock remained practically constant between 1969 and 1974 (Fig. 3.5(e)). The combination of these two trends suggests increasing seed mortality through the period.

Two other increases in mortality also occurred. Firstly, seedling mortality increased, slightly in percentage terms, but significantly in relation to recruitment. Secondly, mortality began to take a heavier toll among older individuals as well (Fig. 3.5(c)). These trends in mortality, survival and fecundity illustrate quite effectively how dynamic processes bring about local spatial and temporal (age-structured) distributions of individuals.

Perennial grasses growing in non-successional communities may also show regular recruitment from seed. Survivorship curves for several such populations have been constructed from annual or more irregular maps of seedlings and adult plant clumps in arid range grassland in the USA and Australia where an interest in the demography of these plants arises from the need to preserve vegetation cover. Comparing the survivorship of eleven perennial grasses growing in semi-desert in southern Arizona, Canfield (1957) found that mortality was increased and lifespan decreased when 'primary' species such as hairy grama *Bouteloua hirsuta* were grazed and that primary species were commonest in ungrazed or 'properly managed' (i.e. not overgrazed) pastures (Fig. 3.6).

Certain 'secondary' grass species such as rothrock grama (*B. rothrockii*) showed the reverse survival responses and distribution under grazing. These grasses' more compact growth form made them less susceptible to severe defoliation by cattle than the tall primary grasses, but their survival was probably also enhanced in grazed conditions because competition from primary grasses would be reduced under grazing.

Predation may reduce plant density and consequently reduce mortality caused by density stress in a number of situations. Seed predation by rodents was observed to have this effect in populations of annual grasses (Ch. 2, p. 28) and in experimental monocultures of shepherd's purse (*Capsella bursa-pastoris*) grazed by slugs (Dirzo and Harper 1980).

Autumn grazing has also been found to increase the survival of Indian ricegrass (*Oryzopsis hymenoides*) and several other grasses of sagebrush-grass communities in southeastern Idaho. West, Rea and Harniss (1979) explained this phenomenon on the grounds that these species are winter dormant so that grazers remove litter rather than green leaf when grazing at this time of year. This may stimulate later growth. Spring grazing of *Oryzopsis* would certainly decrease its survivorship.

Fig. 3.6 The survivorship of some range grasses in grazed populations and ungrazed populations. Date: *Bouteloua* spp. (Canfield 1957); *Oryzopsis hymenoides* (West, Rea and Harniss 1979); *Danthonia caespitosa* (Williams 1970).

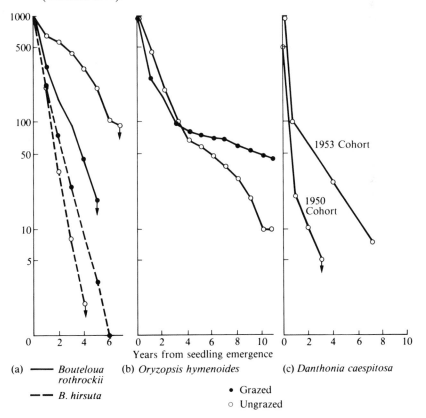

(a) —— *Bouteloua rothrockii*
  – – *B. hirsuta*

(b) *Oryzopsis hymenoides*

(c) *Danthonia caespitosa*

• Grazed
○ Ungrazed

Other factors may also alter plant survivorship. Williams (1970) charted plants of an arid grassland species *Danthonia caespitosa* in Australia and found that cohorts arising in different years had differently shaped survivorship curves (Fig. 3.6). These differences in survivorship are reminiscent of the long-term consequences which emergence order and initial rosette size were found to have on the survival of cohorts of *Viola blanda* and other plants (Ch. 2, p. 34). It is possible that the 1950 and 1953 cohorts of *Danthonia* were born into an environment sufficiently different to alter the initial growth size distribution of early survivors and hence also their later survivorship.

All these grass populations have fairly short half-lives of about 1–3 years and regular (often annual) recruitment from seed. This contrasts with the survival and recruitment of several perennial meadow and woodland herbs studied by Tamm (1956, 1972a, b) in Sweden.

In Tamm's initial study, populations consisted of plants of unknown age. They probably represented the accumulated survivors of many different years' seedling cohorts. Because of this uneven initial age structure it is inappropriate to describe the survival of such a population as survivorship, a term normally reserved for populations which fit the conventions of the life table. Nevertheless, the populations may be treated in a manner analogous to a proper cohort. The 'survivorship' curve for such a population is referred to as a depletion curve to distinguish it from true survivorship curves.

The half-lives of depletion curves, which are often exponential, are inevitably greater than they would be for the survivorship curves of true cohorts followed from the seedling stage because observations are begun with a set of 'proven' survivors. To be anthropomorphic: old heros never die, they just fade away.

Individual plants (probably mostly genets) of cowslip (*Primula veris*) mapped over a period of 29 years at one site had a depletion-curve half-life of 50 years, during which time no seedlings were able to establish and flower. The depletion curve for a population at another woodland site had a similar half-life for the first 14 years of the study but the population then abruptly nosedived with a half-life of 2.9 years when the tree canopy overhead became denser. The percentage of flowering rosettes also declined with age in this population although this trend was not observed in the more stable population. Depletion curves for some orchid populations studied by Tamm are shown in Fig. 3.7. Recruitment to the orchid populations in Tamm's plots was intermittent. This was a characteristic of many of the species he studied. Three other generalizations emerge from his work: 1. survival varies from species to species; 2. survival within a species varies from site to site (*P. veris*); 3. survival within a species at a particular site varies from time to time due to changing conditions.

The age of a number of perennial herbs can be determined from morphological indicators which mark annual growth increments. Thus although no dynamic life table may be available to indicate the frequency with which new recruits enter a population, stage structures or age structures may still provide some information of this sort retrospectively. The assumptions which allow static life tables to be built from age or stage structures are discussed further on page 58.

Bulbiferous species generally display a variety of life histories including regular recruitment. Of three woodland species of wild garlic studied by Kawano and Nagi (1975) in Japan one, *Allium victorialis*, displayed regular annual recruitment from seedlings, while two others possessed stage structures influenced predominantly by recruitment from vegetatively produced bulbils. In one of these species. *A. monanthum*, this produced a stage structure with many juvenile ramets. A stage or age structure for this population based upon genets would undoubtedly appear less youthful.

The flux of individuals through populations is of vital importance, particularly if rare species which appear to exist near the limits of extinction are to be properly managed and conserved.

Fig. 3.7 Depletion curves for some orchid populations. (Data from Tamm 1972)

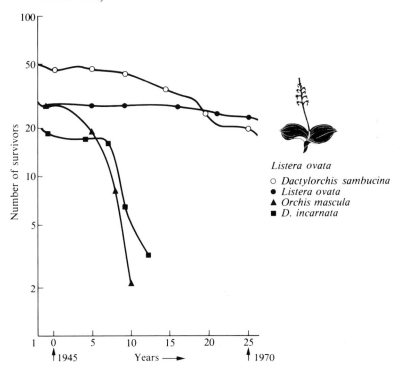

Listera ovata

○ *Dactylorchis sambucina*
● *Listera ovata*
▲ *Orchis mascula*
■ *D. incarnata*

Bradshaw and Doody (1978a, b) monitored population changes in a number of rare species growing as glacial relicts in Upper Teesdale in northern England. Rapid population turnover was found in several species including spring gentian (*Gentiana verna*) which showed regular seed production, and bitter milkwort (*Polygala amara*) in which there was regular recruitment of seedlings (Table 3.1).

Little is known about the relative population flux of plants in sites where they are rare as compared to sites where they are common. In one case where this has been measured for two such contrasting populations of sweet vernal grass (*Anthoxanthum odoratum*), a higher turnover rate of individuals (75% per year) was found in a marginal low-density population than in a more central, dense population (42% turnover per year). The size of the marginal population remained practically constant over the study period of 3 years despite its rapid turnover of individuals (Grant and Antonovics 1978).

One of the first, and still one of the most thorough studies of population dynamics in herb populations is the one conducted by Sarukhan on buttercups, already referred to in Chapter 2.

A comparative study of three buttercups

The object of Sarukhan's study was to examine the flux of individuals through populations of *Ranunculus bulbosus*, *R. acris* and *R. repens* which occurred in an old pasture near Bangor in North Wales. As life tables and fecundity schedules for plant populations accumulate with increasing research in plant population ecology, the novelty of this simple objective will probably seem strange to students of the subject. However, just imagine being unable to answer the deceptively trivial question: 'How long does a buttercup live?' With the wisdom of hindsight the attentive reader would now reply: 'Do you mean a buttercup ramet or genet?' The question is actually quite a complex one whose general and practical applications have already been emphasized.

Table 3.1 Quantitative values of some components of the life strategies of Teesdale plants. (Figures in columns 2–6 are annual rates per 100 plants).

	Half-life (yrs)	Mortality		Seed production	Recruitment	
		Over 1y	Under 1y		Seedlings	Ramets
Linum catharticum	—	100	76	2556		None
Draba incana	1.5	38	63	2541	157	None
Polygala canarella	2.0	31	29	377	54	None
Viola rupestris	8.2	14	35	73	13	16
Viola riviniana	7.2	13	20	56	8	16
Carex ericetorum	1.9	26	26	—	None obs.	36
Gentiana verna	1.8	30	—	1183	None obs.	38

From Bradshaw and Doody (1978)

Since Sarukhan completed his study, a start has been made on the demography of mosses (Collins 1976; Watson 1979), marine algae (Gunnill 1980) and pteridophytes (Duckett and Duckett 1980).

Populations of the three buttercup species were monitored over a period of $2\frac{1}{2}$ years by mapping seedlings and rosettes occurring in permanent quadrats at intervals of a few weeks (Sarukhan and Harper, 1973).

Rosettes of *R. repens* produce daughter rosettes (ramets) at the end of stolons, 20 cm or longer, which later wither, severing the connection between mother and daughter. *Ranunculus repens* also produces some seeds. *Ranunculus acris* produces seeds and also propagates vegetatively but produces fewer daughters than *R. repens*, and these grow on a stolon only a few millimetres from the parent. *Ranunculus bulbosus* produces seeds but has no vegetative propagation. Although it possesses a corm which is renewed annually, it is rare for a rosette to produce more than a single corm at a time.

The maps of these *Ranunculus* populations were used to follow the fate of over 9000 individual seedlings and rosettes and to draw up curves of the cumulative mortalities in populations from the beginning of the study in April 1969 (week 0), Fig. 3.8. All three species showed considerable turnover of rosette numbers. Further analysis of the population changes in *R. repens* showed that patterns of mortality were different for ramets and rosettes originating from seed (genets). The mortality rate in a cohort of ramets formed in summer 1969 was constant and fitted an exponential curve. Few seedlings of *R. repens* emerged simultaneously at any time during the study, and so a survivorship curve for genets was drawn up for an artificial cohort created by lumping together seedlings originating at different dates.

This curve was for a relatively small cohort of seedlings but, in contrast to ramet survivorship, suggested a decreasing risk of mortality with age. A similar pattern of age-dependent survival was also observed for genets of *R. acris* and *R. bulbosus* in early life but survivorship of genets was exponential after about 25 weeks from germination.

To extend the analysis of population changes in *Ranunculus* beyond simple description, Sarukhan and Gadgil (1974) constructed matrix models for populations of the three species from parameters derived from the field studies. A general transition matrix for these populations which contains all the relevant stages of the life histories and the transitions between them is:

		S	R	A	F	G
Seed	S	a_{SS}	0	0	a_{FS}	a_{GS}
Ramet	R	0	0	0	0	a_{GR}
Non-flowering adult	A	a_{SA}	0	0	0	0
Flowering adult	F	0	a_{RF}	a_{AF}	0	0
Flowering and ramet producing adult	G	0	0	0	a_{FG}	a_{GG}

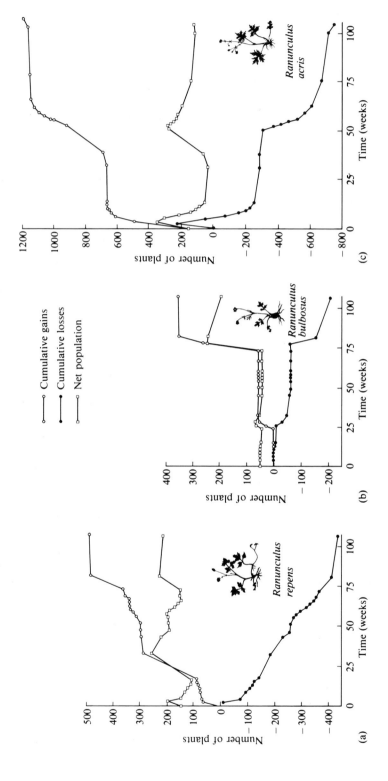

Fig. 3.8 Population flux in *Ranunculus* species. (From Sarukhan and Harper 1973)

Transition probabilities between the S, R, A, F and G stages of the life cycles varied with the seasons and consequently five different transitions matrices were considered necessary for each species, representing spring, early summer, late summer, autumn and winter. The precise dates of each 'season' were adjusted for each species so that certain transitions such as seed germination could be conveniently restricted to a single interval.

Since the different species had different values for the transition probabilities and, as in the case of *R. bulbosus*, some additional elements such as a_{GR} were zero because the species does not produce ramets, these models could be used to test the relative importance for population increase of various modes of reproduction in the different species. This was done by comparing the net reproductive rates R_o achieved by iterating transition matrices in which ramet or genet production was increased or decreased relative to the average performance of the population (Fig. 3.9). As is to be expected, population growth in *R. bulbosus* is highly sensitive to changes in its sexual reproduction since this is its only means of population increase. *Ranunculus repens* is almost as sensitive to increases in *ramet* production but is little affected by increases in genet production as high as threefold above average. When *ramet* production by *R. repens* is reduced below average, R_o is only slightly reduced because ramet mortality is density dependent (Ch. 5 p. 109). Population growth for *R. acris* is more strongly affected by increases in sexual reproduction than ramet production.

Sarukhan's censuses of the *R. repens* population were continued by Soane. Soane and Watkinson (1979) used 4 years' accumulated data to examine genet turnover and recruitment in this species in more detail than was possible within the more limited period of study. They found that genets (families of ramets) had approximately exponential survivorship. After 4 years the families surviving from the first year made a large contribution to the maintenance of the ramet population but a relatively small contribution to the number of different genets present. Though few in number, seedlings recruited since the first year made a significantly greater contribution to the genetic diversity of the population then the oldest, most prolifically ramifying genets.

The demography of shrubs

Few shrub species have been studied demographically because they are neither short lived enough to provide convenient material for the population ecologist nor economically important enough to attract the interest of foresters. Consequently, dynamic life tables only exist for three or four species. The survivorship curves for small cohorts of two of these, *Acacia burkittii* (Crisp and Lange 1976) and *Artemisia tripartita*

(West, Rea and Harniss 1979) show a Deevey type III survivorship (Fig. 3.10). Both are plants of arid rangelands, a habitat in which shrubs are frequently important constituents of the sparse vegetation.

A cultivated population of the temperate shrub broom (*Sarothamnus scoparius*) has been studied in some detail by Waloff and Richards (1977) in Berkshire, England. They transplanted 240 broom seedlings into two plots in which their mortality and natality (seed production) were followed for 11 years. The plants in one plot received a regular insecticide treatment to the foliage and surrounding soil which substantially reduced the populations of phytophagous insects on them. Thirty

Fig. 3.9 Theoretical rates of increase for the three *Ranunculus* species based upon varying modes of reproduction and reproductive parameters. (From Sarukhan and Gadgil 1974)

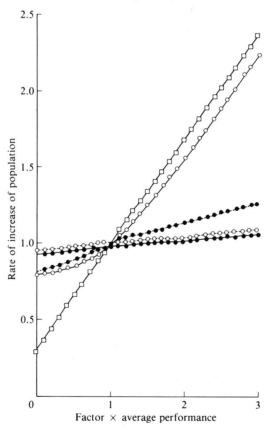

o–o *Ranunculus repens*, sexual reproduction modified
o—o *R. repens*, vegetative propagation modified
•–• *R. acris*, sexual reproduction modified
•—• *R. acris*, vegetative propagation modified
□–□ *R. bulbosus*, sexual reproduction modified

species of phytophagous insects attacked broom in the unsprayed plot, including aphids on leaves and seed pods, lepidopteran stem-miners, seed weevils, agromyzid leaf-miners, beetles feeding on the flowers and a weevil whose larvae feed on root nodules and whose adults feed on broom leaves. Under pressure from this onslaught shrubs in the unsprayed plot suffered twice the mortality of those in the sprayed plot (Fig. 3.10), and in 10 years produced only 25 per cent of the seeds per

Fig. 3.10 Survivorship curves for three shrub populations. Data: *Sarothamnus scoparius* (Waloff and Richards 1977); *Acacia burkittii* (Crisp and Lange 1976); *Artemisia tripartita* (West, Rea and Harniss 1979).

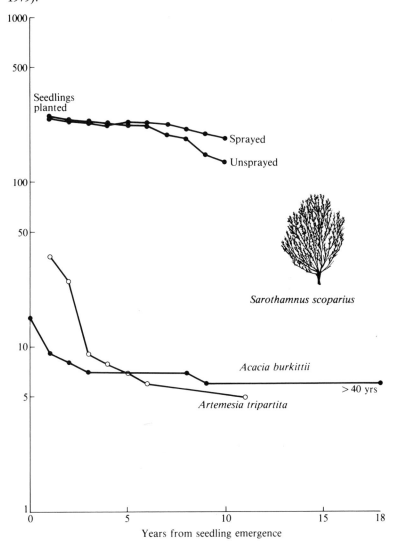

bush produced by sprayed shrubs. The increased mortality caused by insects in the untreated population appears only to have accelerated the mortality which began to take effect at about 8 years in the sprayed population. Senescence in all these plants was associated with a halving in the ratio of weight of green tissue (twigs and leaves) to wood during the 10-year study period, though natality per bush did not decline significantly with age. A population model for broom including the effects of phytophagous insects on mortality and natality would produce instructive results but should include data (which are unfortunately lacking in this study) on the survival of seeds in the seed pool and survival during the first year of seedling life. These data might easily be collected.

The practical difficulties plant demographers face when confronted with plant populations in which individuals live longer than the investigator are not insuperable. It is fortunate that many (though not all) of the longer-lived species can be aged from the annual growth increments (rings) laid down in the woody stem. Various refinements have been devised for obtaining the most accurate estimate of age by dendrochronology (Roughton 1962). The end-product of all methods is a population age structure, equivalent to the column matrix in a matrix model (see p. 8). An example of such an age structure for *Acacia burkittii* is shown in Fig. 3.11. Having obtained a population age structure, the next problem is one of interpretation. A population age structure gives us a glimpse of a single frame in a movie film. Only inference can tell us what happened in earlier frames and only patience or extrapolation can tell us what follows.

Any given age structure can be reconstructed if we assume the correct age-specific mortality and natality that operated during the earlier part of the movie. Taking a simple hypothetical age structure as an example, the pattern in Fig. 3.12 can be explained with the following assump-

Fig. 3.11 Age structures for two contrasting populations of *Acacia burkittii*. (Crisp and Lange 1976)

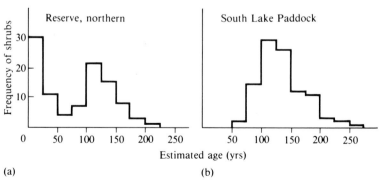

(a) (b)

tions: 1. natality was constant each year; 2. mortality varied each year; 3. mortality acted on first-year seedlings only.

Fig. 3.12 (*See text*)

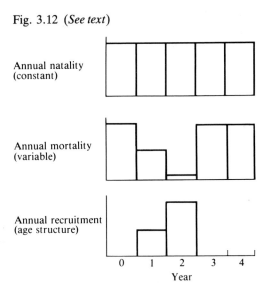

Equally, the age structure of our hypothetical population could have been produced by variable annual natality and a constant, low annual mortality:

Fig. 3.13 (*See text*)

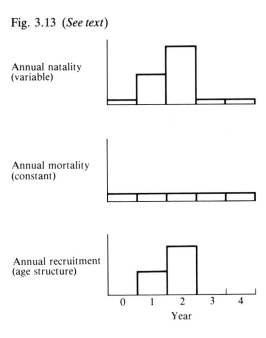

There are, of course, combinations of varying natality and varying mortality between the extremes we have illustrated which could also produce the observed age distribution.

These models assume that the number of individuals now 5 years old is determined by the number born 5 years ago minus the number of those that died in that year. It makes no allowance for those in the original cohort that died in their second or third year. This may be a more reasonable simplification than at first appears because some, probably many, plants do have type III survivorship and massive juvenile mortality. How reasonable is the other assumption about seedling mortality, that it has been very variable in the history of the population? The answer to this question must, of course, depend on the habitat but there are many cases of peaks in population age structures coinciding with historically dated episodes such as a reduction in grazing or the incidence of fire. The dearth of *Acacia* plants younger than 100 years in Fig. 3.11(b) (p. 58) coincides with the introduction of sheep to the area. About 50 years before these shrubs were aged the population in Fig. 3.11(a) was fenced and a new peak is seen in its age structure. The population of Fig. 3.11(b) was left unfenced.

The truncation of the older age classes in the distributions are presumably simply due to deaths from old age in this case. Fire is a common source of disturbance in many shrub habitats and is an event to which many species appear resistant. The seeds of ling (*Calluna vulgaris*) which occurs on British and northwest European heaths germinate most effectively after a brief burning of the litter on the surface of the soil. Whole woodland communities in Western Australia appear to depend upon fire for effective regeneration, not only because seed germination in many species has become fire dependent but also because fires purge large areas of the phytophagous insects which consume any herb, shrub and tree seedlings which appear in unburnt areas (Whelan and Main 1979). A fire of this kind will obviously produce sharp peaks in population age structures.

Variations in mortality and natality may produce peaks in age structures by truncating younger age classes. What would the age

Fig. 3.14 (*See text*)

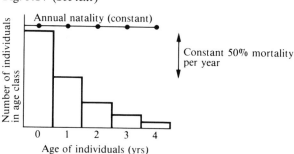

structure of a population with constant annual natality and some constant annual value of mortality (say 50%) look like?

Assuming a constant yearly input of seedlings to the 0 age class, a constant percentage mortality per year would produce an age structure with an exponential profile (Fig. 3.14). In fact the assumption of constant annual natality enables us to treat a static life table based on age structure exactly as a dynamic life table. A histogram of a cohort of seedlings with exponential survivorship drawn from a dynamic life table would look just the same as Fig. 3.14.

Again we must ask, how well does the assumption of constant annual recruitment fit the facts? Obviously in some cases such as in fire-prone ecosystems where germination is dependent upon fire, recruitment is highly variable between years. On the other hand, information collected by foresters in North America on the seed production frequency of shrubs shows that such species vary far less in their year-to-year reproduction than do trees (Silvertown 1980b). The age structure of a population of the antelope bitterbush (*Purshia tridentata*) growing in Colorado (Fig. 3.15) is probably a product of regular natality combined with a pattern of declining mortality with age such as that observed in the dynamic study of *Artemisia tripartita* (Fig. 3.10, p. 57). An age structure indicative of an actively regenerating population appears to be less usual for shrubs than age structures with irregular peaks reflecting isolated, strong cohorts produced by irregular recruitment (Roughton 1972).

Complications arise in the ageing of populations by dendrochronology when the plants are clonal. Many shrubs show the clonal habit which may be favoured in habitats where the chances of successful seedling establishment are extremely low. Bilberry (*Vaccinium myrtillus*) and bearberry (*Arctostaphyllos uva ursi*) are examples found at high altitudes in northern Europe. Shoot populations of *V. myrtillus* which

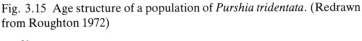

Fig. 3.15 Age structure of a population of *Purshia tridentata*. (Redrawn from Roughton 1972)

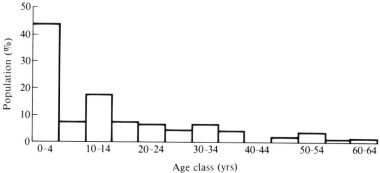

have been aged in Scotland and in Sweden (Flower-Ellis 1971) have an age structure that suggests regular recruitment (Fig. 3.16). However, the shoots aged in this population are merely the latest cohorts of ramets produced by rhizome systems of unknown ancestry. This is a somewhat similar situation to that found in *Carex bigelowii* and other arctic species (p. 12).

The creosote bush (*Larrea tridentata*) is another clonal shrub, widespread and often dominant in desert areas of the Southwest USA and northern Mexico. The age structure of a population of genets expanding into a new area in southern Arizona showed a peak of 15–20-year-old shrubs with little subsequent recruitment (Chew and Chew 1965). Observations of the clonal extension of *Larrea* bushes in the Mojave desert suggest that some genets there may be thousands of years old. Central stems die and clonal extension forms a ring of shrubs (ramets) which advance radially at a rate of less than 1 mm a year. Extrapolating from this modern rate of extension, Vasek (1980) calculated that the largest clone in his study area could be 11 700 years old. An invasion of the kind observed in Arizona could therefore be a major founding event in the genetic history of a local *Larrea* population.

Fig. 3.16 Survivorship curves for cohorts of *Vaccinium myrtillus* ramets at five sites. Based on a static life table derived from ramet age structures. (From Flower-Ellis 1971)

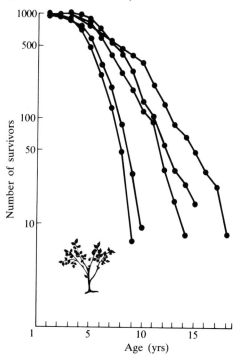

The demography of trees

Though the demography of trees has received more attention than that of shrubs, this is not reflected in the number of dynamic life tables available for trees. Instead it seems only to have increased the number of times foresters and ecologists have remarked on the apparent irregularity of recruitment to tree populations. On reflection it is not surprising that most of the millions of seeds an oak produces in an average reproductive lifespan perish from one cause or another before they reach maturity. On average only one acorn from a lifetime's crop has to survive to maintain an equilibrium population size.

The seed production of many forest tree species is periodic, with occasional large crops interspersed by years in which few or no seeds at all are produced. Conspecific trees in the same geographical area that show this habit, which is known as masting, usually have synchronized reproduction and consequently there may be very large differences from year to year in the number of seeds reaching the forest floor. Studies of the seed pool in both temperate and tropical primary forests (Ch. 2) show that few tree seeds become incorporated in the soil, and among tropical trees in particular, germination usually takes place soon after dispersal (Ng 1977). Masting and rapid germination are thought to be defensive strategies against pre-dispersal and post-dispersal seed predation respectively. There is strong evidence that smaller proportions of large crops than of small ones fall prey to insect and vertebrate seed predators because predators are swamped by an excess of food in mast years. (Ch. 4).

When the seed production of individual trees is measured over the entire lifespan, with the between-year variations caused by masting averaged out, the fecundity schedule often shows an increase in seed production with size and age. The fecundity schedules for three oak species measured in the southern Appalachians by Downs (1944) and for three neo-tropical trees all show such an increase in varying ways (Fig. 3.17). The decline in seed production with size (and presumably also with age) in *Quercus alba* and *Q. rubra* also occurs in fruit trees. It would probably also be observed in other species if the oldest age classes were studied.

In fact the number of seeds a tree produces gives us no clue to the importance of reproduction by seed in its population dynamics. The coastal redwood (*Sequoia sempervirens*) may produce seeds annually but it also produces vegetative sprouts from the stumps of felled trees. The aspens *Populus tremuloides* and *P. grandidentata* are also clonal (Barnes 1966), as is the English elm (*Ulmus procera*). This species produces an abundance of seed in Britain, most of which is non-viable and its local distribution (until recent devastation by elm disease) must largely be due to vegetative spread. Black poplar (*P. nigra*) is a

Fig. 3.17 Fecundity schedules for six tree species. Data: *Quercus* spp. (Downs and McQuilkin 1944); *Pentaclethra macroloba*, *Welfia gorgii* and *Astrocaryum mexicanum* (Sarukhan 1980, after various authors).

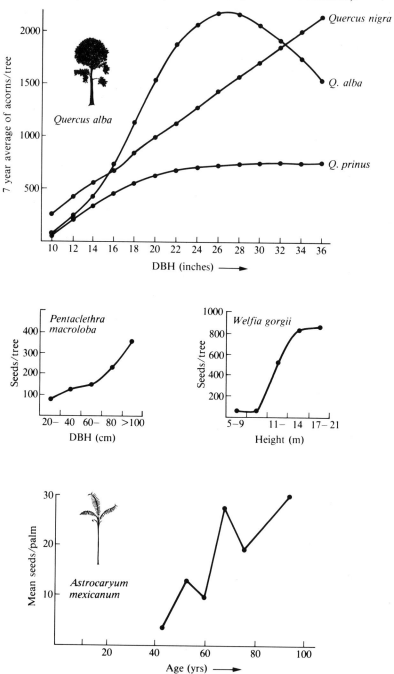

dioecious vegetative species, common in East Anglia during the Middle Ages but now possibly represented in Britain by under 1000 individual trees, most of them male (Rackham 1976).

The fate of seeds before and after dispersal is obviously of crucial importance in determining the potential recruitment from seed. All indications are that loss through animal predation in most populations is enormous. Birds and small mammals were found to have destroyed 69 per cent of the seeds of Douglas fir (*Pseudotsuga menziesii*) falling into a clear-cut in Oregon (Gashwiler 1967). Vertebrate predators took 98.5 per cent of the acorns produced by sessile oak (*Quercus petraea*) in a Welsh woodland studied by Shaw (1968), and bruchid beetles have been found to account for the loss of up to 100 per cent of the seed crops of various legume trees in Panama (Janzen 1969). Fruits of the understorey palm *Astrocaryum mexicanum* which has been studied by Sarukhan (1977, 1980) in high evergreen tropical forest in Mexico have only a 5 per cent chance of reaching the germination stage alive. Squirrels account for 50 per cent of this loss, reaching the fruit while it is still on the tree by the most roundabout route to overcome the palm's defences:

> The squirrel avoids the numerous spines on the trunk and underneath the leaves by climbing from a nearby branch to the top of the palm crown, and sliding on the upper sides of the leaves (which are unarmed) until these bend low under the weight of the animal. In this position, the squirrel can reach the hanging infructesences without touching either the trunk or the basal parts of the petioles, which are heavily armed with spines (Sarukhan 1977:177).

Fig. 3.18 Survivorship curve for the palm *Euterpe globosa*. (From Van Valen 1975)

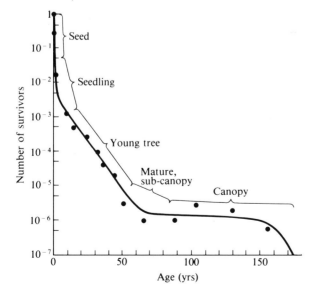

Seedling and particularly seed mortality is so great for most trees that a full survivorship curve may span eight orders of magnitude. Such comprehensive curves are seldom plotted. An unusual case in which mortality at every stage of the life cycle has been estimated (mostly from population stage structure and a static life table) is plotted on a single diagram in Fig. 3.18. Seedling survivorship curves drawn from dynamic life tables for other tropical and temperate trees show patterns of mortality similar to that of *Euterpe globosa* in the first decade of life (Fig. 3.19).

The three species illustrated in this figure come from forests on three different continents but share similar survivorship and appear to be typical of other tree species studied in the same regions (Wyatt Smith 1958; Hett and Loucks 1976; Sarukhan 1980). Tropical dipterocarp seedlings (*Shorea* and other genera) and a number of temperate trees including *Acer saccharum* may persist as oskars for many years in shade on the forest floor (Ch. 2). Considering that these conditions must place

Fig. 3.19 Seed and seedling survivorship curves for three tree species. Data: *Acer saccharum* (Curtis 1957 and Hett 1971; *Shorea parviflora* (Wyatt Smith 1958); *Pinus sylvestris* (Guittet and Laberche 1974).

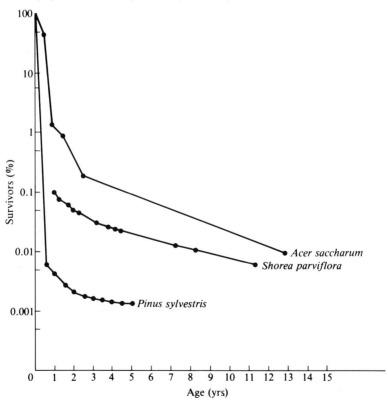

them on the verge of their photosynthetic compensation point, it is surprising that plants in this vulnerable situation show declining death-rates with age.

Static life tables for *A. saccharum* have also been compiled by Hett and Loucks (1971) who fitted two kinds of mathematical relation to the age structures of eight populations to test whether estimated survivorship conformed to Deevey's type II or type III. The exponential equation:

$$N_x = N_0\,e^{-bx}$$

or in natural log form:

$$\ln N_x = \ln N_0 - bx$$

describes type II survivorship and a power function can be used to describe type III:

$$N_x = N_0 x^{-b}$$

in log form:

$$\ln N_x = \ln N_0 - b\ln x$$

where N_0 is the number of seedlings at time zero, N_x is the number at time x and b is the slope of the curve and hence the mortality rate. The power function was found to give a better fit to the eight age structures than the exponential model and pointed to a declining mortality rate with age in these populations. Of course the fact that this conclusion coincides with the dynamic analysis of mortality for this species in Wisconsin is further evidence that age-structure analysis can yield reliable information about survivorship.

Similarly shaped age structures, implying type III survivorship, have been identified in many forest tree populations including striped maple (*A. pensylvanicum*) (Hibbs 1979), balsam fir (*Abies balsamea*) and eastern hemlock (*Tsuga canadensis*) (Hett and Loucks 1976) in North American forests; Chile pines (*Araucaria* spp.) in New Guinea (Gray 1975) and three neo-tropical trees *Astrocaryum mexicanum*, *Euterpe globosa* and *Pentaclethra macroloba*. This list could be expanded considerably.

Age structures of this type are frequently referred to by foresters as 'reverse J' in shape, and it is often said that such an age structure represents a stable age distribution. Although this may be true in certain circumstances, strictly speaking it cannot usually be said that the age structure of the population will undergo no future changes and it is safer to say only that regeneration is taking place, thus making a shift towards a 'geriatric' age structure and all that this implies unlikely.

The causes of tree mortality are of course of vital interest to foresters and are well documented (Kulman 1971; Gray 1972). The diseases and insects which attack trees are often age specific. Seedlings are prone to fungal 'damping-off' disease for instance, while defoliation of tamarack

(*Larix larcina*) by the larch bud moth may affect mature trees in a reproductive state more severely than juveniles (Niemela, Tuomi and Haukioja 1980). Other diseases attack trees already weakened by storms or insect attack. Moth caterpillars are often responsible for the severe defoliation of trees. Insects such as the winter moth, gypsy moth and the spruce budworm experience population explosions in certain years. Defoliated trees show a reduced increment in growth but most can recover from a single defoliation, though repeated attacks are increasingly likely to be fatal (Kulman 1971).

Certain insects associated with trees are protective. Oak, birch and conifers near mounds of the wood ant (*Formica* spp.) are not defoliated during moth and sawfly outbreaks which strip the trees found further away from these mounds. In a study of birch (*Betula pubescens*) in northern Finland it was found that trees within 30 m of ant mounds showed fewer signs of leaf predation even in a year of normal herbivore density (Laine and Niemela 1980). The relationship between ants and trees has evolved into a highly developed symbiosis in the ant acacia which is protected from insect predation and from the interference of other plants by the ants which it harbours in special thorns (Janzen 1973).

Data on the age-specific causes of mortality in the life cycle of western white pine (*Pinus monticola*) were collated by Waters (1969) from observations made in North American forest plantations of this species. The quantitative impact of the various mortality factors was used to compile an artificial dynamic life table which is shown graphically in Fig. 3.20. This life table is artificial in two ways: firstly, it is the product of observations made on several, different, even-aged populations in order to obtain data on trees of different ages; secondly, not all the

Fig. 3.20 Survivorship and the causes of mortality for a model cohort of *Pinus monticola*. (Drawn from data of Waters 1969)

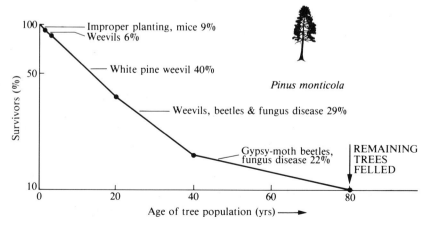

'mortalities' recorded are genuine deaths because Waters was primarily interested in the economic losses caused by disease in forest stands and so he recorded some trees which were rendered economically useless by disease but which were not actually dead as fatalities. The information is still of demographic value since morbidity, particularly in a dense population, is often a prelude to mortality. As in the study of planted seedlings of broom (p. 56), the life table for pine plantations underestimates the mortality which occurs in most natural populations in the earliest months and years of seedling establishment.

While forest populations with reverse J age structures are common, so too are populations which show a strong peak of ageing trees with little or no recent recruitment. For instance, age distributions indicating increasing, static or declining populations are found according to which species is examined in virgin conifer and broadleaf forests in New Hampshire (Leak 1975). Clearly the species composition of these forests, as well as the age structure of some of their tree populations, are changing. Such changes appear to be typical of most North American and other forests, whether 'virgin' or not (P.S. White 1979).

Foresters often distinguish between species they call 'shade tolerant' which show a capacity for regeneration in shade by a reverse J age structure, and species called 'light demanding' which have no young trees in the understorey. Of course there are species which fall between these two extremes. Age-structure analysis of over 800 stems of black cherry (*Prunus serotina*) in oak forest in southern Wisconsin by Auclair and Cottam (1971) showed that 58 per cent of these were between 20 and 30 years old. They attributed this pattern to the effect of disturbance in the forest during the 1930s when farmers cut foliage from the forest to feed their cattle. This opening of the forest canopy must have 'released' some suppressed seedlings and enhanced the survival of new seedlings through the most vulnerable part of the life cycle. Gaps in the forest conopy have a similar effect on populations of pin cherry (*P. pensylvanica*) which produces an abundance of new seedlings from dormant seeds in the soil where trees are felled (Marks 1974). Both these species of cherry are dependent on gaps in the forest canopy for regeneration, and in unmanaged woodland would depend upon natural tree fall and windthrows to permit regeneration from seed (*P. pensylvanica*) or from seed and persistent oskars. (*P. serotina*).

Gaps in the forest canopy appear to be of universal importance for the regeneration of many species in both temperate and tropical, managed and unmanaged woodlands. An estimated 75 per cent of the 105 tree species composing the canopy of a primary tropical wet forest studied in Costa Rica depend upon canopy gaps for successful regeneration (Hartshorn 1977). Such gaps are frequently created by tree falls in the forest Hartshorn studied, opening up 1 per cent or more of the ground area annually. It would take only 118 ± 27 years for complete turnover of the trees in this forest. This is a far greater rate of turnover than is observed in temperate forests, such as maple–hemlock forest in Wisconsin (Stearns 1949) or in the New Forest in England, but it may be representative of tropical forests.

Fire is a common source of natural disturbance in many forests. Its role is best documented in North America where conifer species in particular appear to depend upon it for regeneration. A natural 'fire rotation' of about 100 years is thought to have prevailed in pine forests in Minnesota before settlement times (Heinselman 1973; Swain 1973). Mature trees of forests in the Sierra Nevada, California, the northern Rocky mountains, Grand Teton and Yellowstone National Parks and in the boreal and taiga forests of Alaska, as well as in many other areas, originate from seedlings recruited after fire (Wright and Heinselman 1973). Documentary evidence and fire scars on standing trees have been used to map areas burned between 1712 and 1913 in 130 km^2 of forest in Minnesota. Stands of pine (*Pinus banksiana, P. resinosa, P. strobus*) occurring in this area were all recruited within areas burnt in the year of their origin (Fig. 3.21).

Hurricanes are an important factor in the regeneration of species in many tropical forests. In the first 3 years of a study of changes during a 30-year period in lower montane rainforest in Puerto Rico, Crow (1980) found rapid increases in the number of species and tree stems, reflecting disturbance from hurricanes and forestry a decade before the study began. In 1976, 41 years after the last severe hurricane in the area, these changes had ceased and the forest was approaching a steady state. Of the most abundant species in the forest, only the shade-tolerant palm *Euterpe globosa* increased its population significantly in the final 20 years of the study.

Windthrow is also an important factor in producing gaps in Far Eastern rainforests. In areas prone to cyclones these gaps may be huge and whole areas of forest will begin the process of regeneration simultaneously, with the result that forests of uniform structure and floristic composition develop. In cyclone-free areas, such as the Sunda shelf rainforest of Sumatra, Malaya and Borneo, treefalls create much smaller gaps that are colonized piecemeal, forming a mosaic of vegetation structural types and species (Whitmore 1977). Within a single forest type, gaps of different sizes may be colonized by different sets of species. This was demonstrated in an elegant experiment by Kramer (cited by Whitmore 1977), who made artificial gaps of 0.1 ha, 0.2 ha and 0.3 ha in primary forest in Java. The smallest gaps were rapidly colonized by surviving young individuals of the primary forest species but in larger gaps these were choked by other invading tree species.

In this chapter we have seen how the demographic characteristics of plant populations vary between species and with different habitats. Such characteristics may be seen both as adaptations to life in particular habitats and as constraints which prevent species expanding their populations into some other habitats. We will explore the adaptive element in one important aspect of plant demography–reproduction, in the next chapter. We return to the subject of demographic constraints on plant distribution in Chapter 8.

Summary

The shape of a survivorship curve is influenced by the life history characteristics of a species and by local environmental conditions.

Fig. 3.21 The distribution of fires and recruitment of pines between 1712 and 1913 in Itasca State Park, Minnesota. (From Frissell 1973)

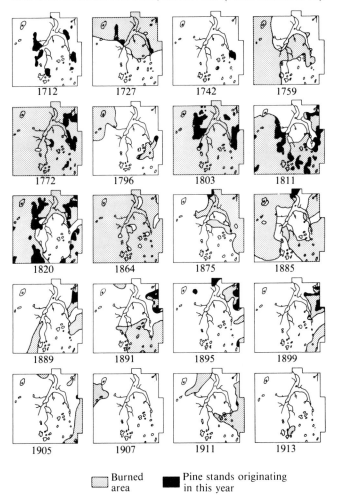

Deevey classified survivorship curves into three general types according to the distribution of mortality risk with age.

With exceptions, annuals often have type I curves, monocarpic perennials type II and other perennials type III curves. A reserve of dormant seeds or regular environmental disturbance are necessary for the persistence of populations of short-lived plants. Recruitment from seed is more intermittent among many herbaceous perennials and trees but environmental disturbance is still important to many of them.

Grazing by stock, insect predation and slug grazing have all been shown to affect plant survivorship, in some cases actually enhancing survival. These factors also affect fecundity which is influenced by individual plant age or size and by population density as well.

4

The ecology of reproduction

In the foregoing chapters we have looked at the demographic character-istics of a range of plant species and have seen how variations in age-specific birth- and death-rates affect the dynamics of populations. Even though relatively few plant populations have been studied, it is clear that there are some crucial differences between species in many aspects of their life histories. Can we make sense of these differences in the reproductive and survival characteristics of species by looking at other aspects of their ecology? To what extent can the average longevity of a plant, the number of times it reproduces and the number and size of its seeds be explained as characteristics which are of some advantage in its typical habitat? In answering these questions we must look at the effects of different reproductive patterns on the fitness of plants displaying them.

Reproduction versus growth

The two fundamental components of fitness are reproduction and survival. A simple measure of these components may be obtained from the lifetime sum of age-specific fecundity and age-specific survivorship: $\Sigma l_x b_x$. The number of seeds produced by a plant, the number of seeds it fathers with the pollen it produces and the proportion of these offspring which survive to reproductive maturity are the factors which determine how many descendants are left by a genotype expressing a particular life history pattern.

It is axiomatic that natural selection favours those genotypes which leave the most descendants. Thus we may define those life history patterns which maximize the number of their surviving offspring under particular ecological conditions as *optimal* for those circumstances. Different life history patterns may be represented by life tables and fecundity schedules with different age-specific patterns of survival and fecundity. The optimal life history among a collection of possible alternatives is therefore the one (or ones) which produces the highest value of $\Sigma l_x b_x$. In other words this is the one with the highest fitness.

Although it might be expected that fitness of an average individual would increase in direct proportion to the number of seeds produced, in reality reproduction incurs a 'cost' in terms of growth and survival.

Plants appear to possess only limited resources which are shared between the competing demands of maintenance, growth and reproduction.

The effect of seed production on the annual growth of temperate-forest trees can be determined by comparing the size of seed crops with the width of annual rings produced in the same and following years. Resources allocated to seed production can produce a decrease in wood growth in beech (*Fagus sylvatica*) for more than 2 years succeeding a large crop (Holmsgaard 1956), though in other species such as Douglas fir (*Pseudotsuga menziesii*), grand fir (*Abies grandis*) and western white pine (*Pinus monticola*), large seed crops only reduce growth in the year of seed production itself (Fig. 4.1) (Eis, Garman and Ebel 1965). Root growth is also affected by reproduction. Even light crops of fruit have been observed to reduce root growth in apple trees.

Though we might expect growth and survival to be related, there is no direct evidence that seed production incurs a cost in fitness by increasing the probability that a tree will actually die. Such an effect has been found in some experimental populations of annual meadow grass (*Poa annua*). Despite its name, this plant frequently lives longer than 1 year and its life history varies both within and between populations. The relationship between the number of inflorescences borne by a plant at 4–5 months of age and the probability that it would survive to 18 months old was calculated for a collection of plants grown from seed under standard conditions by Law (1979). This revealed an inverse relationship between early reproduction and survival. Law also found an

Fig. 4.1 The relationship between cone crop size and annual growth increment for a population of Douglas fir (*Pseudotsuga menziesii*). (From Eis, Garman and Ebel 1965)

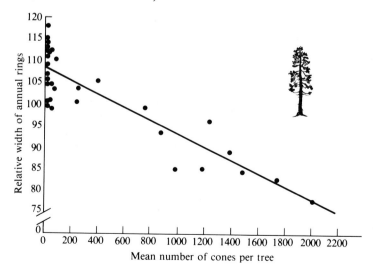

inverse relationship between the number of inflorescences produced by *P. annua* in the first season and both the size of plants and their inflorescence production in the second season.

In longer-lived herbs which spread vegetatively, seed production may exact a cost in terms of reduced clonal growth. The mayapple (*Podophyllum peltatum*) is a rhizomatous, perennial herb which occurs in deciduous woodlands of eastern North America where it was studied by Sohn and Polikansky (1977). The mayapple produces both flowering shoots and new subterranean growth from its rhizome. However, sections of rhizome which produce a flowering shoot and which also bear fruit are generally incapable of producing as much vegetative spread in the following year as rhizomes which have flowers but do not produce fruit.

Mayapple clones are susceptible to various hazards, in particular to a rust disease which prevents growth altogether in the part of the rhizome attached to infected shoots. Sohn and Polikansky calculated the life expectancy of model mayapple clones in which all sexual shoots bore fruit, compared with clones in which some shoots were barren. They found that the probability of eventual extinction for an entirely fertile clone was 1.0 (i.e. certain), but that the probability for a partly barren clone was 0.55. If the environmental causes of mortality did not become more severe than those measured in this study, some partially barren clones could persist indefinitely.

The dichotomous allocation of resources to reproduction or growth may not always be as clear-cut as it appears to be for *Poa* or *Podophyllum*. In many species, reproductive structures and even seeds themselves (Yakovlev and Zhukova 1980) contain chlorophyll which enables them to photosynthesize and hence to make a contribution to the energetic cost of their own production. This energetic contribution may be substantial but reproductive structures do also utilize other resources such as minerals (Bazzaz and Carlson 1979).

The timing of reproduction and death

If reproduction and survival, which includes the chances of further reproduction, are generally alternatives to some extent, then what determines where the balance between these two options lies in a particular population? This question can be broken down into two parts. The first is: what is the optimum length of time for a plant to remain in a non-reproductive phase of growth before producing its first, and for single reproducers its last, seeds and pollen? The second question is: when does a single bout of reproduction followed by death produce a higher fitness than repeated reproduction?

Before following up these questions, it would be advisable to sort out some of the confused terminology associated with this subject. The botanical custom is to divide species into annuals, biennials and perennials which notionally live for 1 year, 2 years and more than 2 years respectively. These terms are not precise enough for our present purpose. Some populations of 'annual' species actually live longer than 1 year (e.g. annual meadow grass). Though most biennials (e.g. *Daucus, Pastinaca, Dipsacus* or *Digitalis*) only flower once, many individuals take longer than 2 years to do so and some such as hoary whitlow grass (*Draba incana*) have populations which are almost entirely perennial (e.g. in Upper Teesdale; Bradshaw and Doody 1978b). A similar spectrum of life history is found in both animals and plants and the same body of theory applies to the explanation of their evolution, so we will adopt a universally applicable terminology.

Semelparous organisms are those which reproduce once and die (referred to in botanical nomenclature as *monocarps*), *iteroparous* organisms reproduce more than once (and are called *polycarps* by botanists). We will use a restricted definition of the botanical terms *annual* and *perennial* to refer to organisms which live for 1 year and more than 1 year, irrespective of how often they reproduce.

Annual plants 'prove' that precocious reproduction is possible and yet many herbs delay reproductive maturity for several years. Some bamboos and some trees may even delay reproduction for several decades. In general terms, precocious reproduction should always be favoured over delayed reproduction in a growing population because of the compounding effect of reproduction. Ten offspring from an annual plant will have 10 offspring each in the following year, these will in turn have

Fig. 4.2 Percentage of the herbaceous flora accounted for by annuals plotted against coefficient of variation in total annual rainfall (CV) in five North American desert habitats. (From Schaffer and Gadgil 1975 after K. T. Harper)

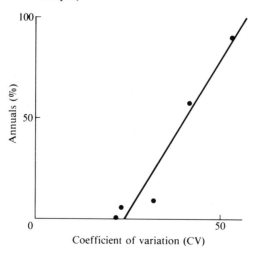

10 offspring each and in 6 years time the original annual plant will have given rise to 10^6 descendants. A plant taking 2 years to reach maturity will have only 10^3 descendants at the end of the same period.

Several different ecological situations favour delayed reproduction, but one consideration is fundamental. Due to the vegetative costs of reproduction, a plant reproducing at small size may experience a greater risk of death. Thus $\Sigma l_x b_x$ may be maximized by delaying reproduction until the plant reaches such a size as to be able to survive, or at least to complete, its first bout of reproduction. In general, the optimum age of reproductive maturity is reached when no further increase in $\Sigma l_x b_x$ can be obtained by any further delay.

Environmental causes of mortality (not directly related to the cost of reproduction) may determine where the upper limit of this optimum age lies. In unpredictable habitats such as deserts, plants which delay reproduction experience a high risk of dying before reproductive maturity. This is a plausible explanation for the high proportion of annuals found in desert floras (Fig. 4.2).

High mortality risks at any stage of the life history favour early reproduction. Law, Bradshaw and Putwain (1977) compared the life histories of *Poa annua* populations collected from disturbed environments such as building sites where mortality risks were high, and from relatively stable pasture environments. Populations of plants grown from seeds collected in these two types of environment were raised in uniform conditions and their age-specific survival (l_x) and reproduction (b_x) were determined at monthly intervals (x) (Fig. 4.3).

Fig. 4.3 Survivorship (dotted line) and fecundity schedules (histogram) for two populations of *Poa annua* derived from contrasting environments. (From Law, Bradshaw and Putwain 1977)

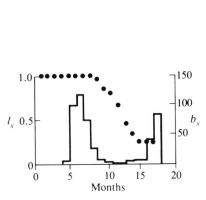

Population from a disturbed environment

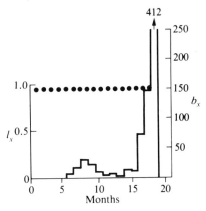

Population from a stable environment

The population from a disturbed environment showed early reproduction and early death, while reproduction in the pasture population
was delayed so that only one major period of reproduction could be
observed in the 18-month duration of the experiment. Although the
short-lived population produced more inflorescences than the long-lived
one in the first reproductive season (5–10 months), the situation was
reversed in the second season (15–18 months). Delayed reproduction
and increased survival in the pasture population allowed increased
growth per plant and an increased average reproductive output.

Even if a small plant devotes all its resources to reproduction, it can
only produce a relatively small crop of seeds. Small crops generally
suffer proportionately more losses to seed predators than large ones
because large crops may swamp animals with more food than they can
handle. This creates a situation in which a plant may gain a dispro-
portionate release from predation by delaying reproduction until it is big
enough to increase its crop size and swamp its seed predators.

This is the explanation which has been put forward for the most
spectacular delays of reproductive maturity which are found in species
of semelparous bamboo, among which a 20-year pre-reproductive
period is common. One species, *Phyllostachys bambusoides*, waits 120
years to flower and die. These bamboos also synchronize reproduction
within cohorts of the same age. This synchrony is probably important
because it reduces the risk that predator populations will move from one
bamboo population to another as they fruit (Janzen 1976).

Some idea of the optimal *frequency of reproduction* can also be
obtained by considering the effect of fecundity and survivorship patterns
on $\Sigma l_x b_x$. Consider a hypothetical population of semelparous annual
plants such as the one shown in Fig. 4.4(a). Plants produce three seeds
each and then die at the end of 1 year. Each seed germinates, there is no
seedling mortality and nine seeds are produced in the second year for
every three present in the previous one. What advantage in fitness
would a mutant individual in this population gain if it became iteroparous? As Fig. 4.4(b) shows, the increased reproduction (equivalent to
increased fitness in this case because the probability of survival = 1) of
such a mutant would be equal to that of a semelparous individual which
produced one extra seed before dying. It seems reasonable to suppose
that the mutant would have to sacrifice more than one seed in order to
switch to iteroparity in the first place so that the situation as it stands in
our model appears to be heavily weighted in favour of semelparous
plants. Cole (1954), who derived this result, pointed out that it
paradoxically predicts that all organisms should be semelparous.

We have plainly left something important out of our model so far. In
fact it is the second component of fitness – survival. Heavy seed and
seedling mortality is commonplace in natural populations and values of
more than 90 per cent for pre-reproductive mortality are not unusual

(Ch. 3). For arithmetical simplicity let us assume a modest pre-reproductive mortality risk of one-third (33%) in our model population of three semelparous individuals (Fig. 4.4(c)). The mean number of seeds produced per original plant in this population is now four, but increases to six if plants which reach reproductive age in the first year do not die but are iteroparous and survive to flower again in following seasons.

Assuming for the moment that the probability of an adult plant surviving is one, then semelparous plants cannot match the advantage in seed production obtained from iteroparity by producing just one additional seed, since each seed produced only has a two-thirds chance of surviving to reproduce. In fact semelparous plants will have to produce one and a half more seeds to match the fitness of iteroparous

Fig. 4.4 (a) A model population of three semelparous plants; (b) A model population of three iteroparous plants; (c) model semelparous population including a one-third mortality risk to seedlings; (d) model iteroparous population including a one-third mortality risk to seedlings. (From Open University 1981)

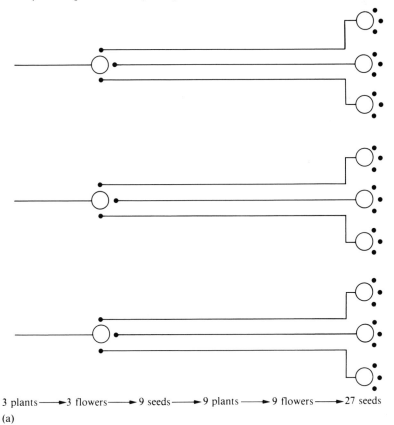

3 plants ⟶ 3 flowers ⟶ 9 seeds ⟶ 9 plants ⟶ 9 flowers ⟶ 27 seeds

(a)

Fig. 4.4b, c

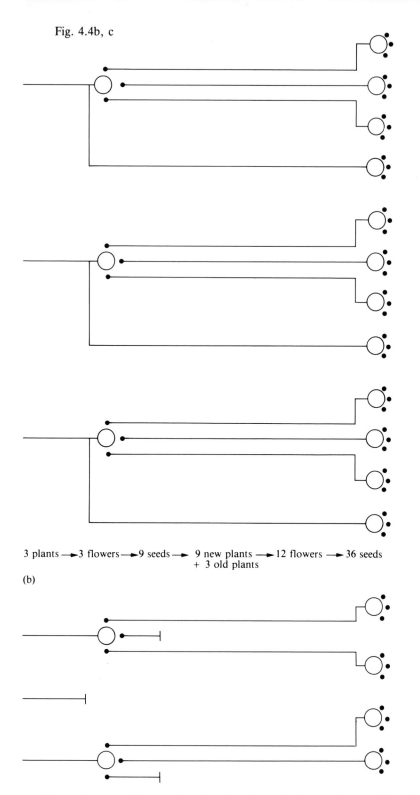

3 plants ──►3 flowers──►9 seeds──► 9 new plants ──►12 flowers ──► 36 seeds
                                              + 3 old plants

(b)

3 seedlings ►2 plants ►2 flowers ► 6 seeds ► 6 seedlings ►4 plants ►4 flowers►12 seeds

(c)

plants and ten more seeds when pre-reproductive mortality is 90 per cent.

This reasoning depends upon the assumption that adult plants have a zero risk of mortality. Though this is clearly untrue, the actual advantage of an iteroparous life history over a semelparous one depends upon the ratio of adult survival (p) to seedling survival (c) (Schaffer and Gadgil 1975). The population of a semelparous annual will increase at a rate R_a, given by the product of annual seed and seedling survival (c) and mean seed production per individual (B_a):

$$R_a = cB_a \qquad [4.1]$$

The annual rate at which an iteroparous perennial population increases (R_p) is given by an expression of the same form plus the mean probability of an adult surviving, p:

$$R_p = cB_p + p \qquad [4.2]$$

A semelparous annual will then reproduce faster than an iteroparous perennial when:

$$B_a > B_p + \frac{p}{c} \qquad [4.3]$$

Sarukhan and Harper (1973) determined values of c, B_p and p for the iteroparous perennial *Ranunculus bulbosus* in the study discussed in Chapter 3 ($c = 0.05$, $B_p \simeq 30$, $p \simeq 0.8$). A semelparous mutant in this

Fig. 4.4d

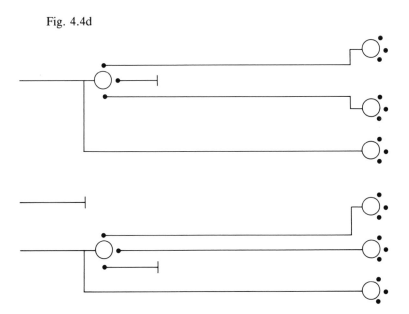

3 seedlings ► 2 plants ► 2 flowers ► 6 seeds ► 6 seedlings ► 4 new ► 6 flowers ► 18 seeds
                                                              + 2 old
(d)   plants

population would have to increase its seed production by 53 per cent in order to match the fitness of iteroparous plants (Schaffer and Gadgil 1975).

The survivorship curves for semelparous and iteroparous populations reviewed in Chapter 3 can now be compared with the predictions of this model. The model predicts that populations which experience relatively heavy juvenile mortality and which have Deevey type III curves should be iteroparous, while populations with shallower curves should tend towards semelparity. The data of Figs. 3.3, 3.4, (semelparous species) and 3.6, 3.11, 3.19, 3.20 (iteroparous species) are not inconsistent with this (see pp. 43–68). The model also suggests that semelparous plants should produce more seeds per crop than iteroparous ones which grow in the same habitat. There is also some evidence to support this (Salisbury 1942).

Reproductive allocation

As we have seen, the balance between reproduction and growth in a whole lifetime can be quantified by measuring the age of reproductive maturity, age-specific seed production and survivorship. We need to make different measurements to quantify the balance which is set between reproduction and growth within a single season. This balance is expressed as the *reproductive allocation* (RA) of a plant which is defined as the proportion of a plant's annual assimilated resources which is devoted to reproduction. In practice it is determined by dividing the dry weight of plants into the weight of reproductive and non-reproductive parts.

An example of how assimilated carbon is portioned between various plant organs in groundsel (*Senecio vulgaris*), an annual composite, is illustrated in Fig. 4.5. Although the total net assimilation and total seed production may be decreased drastically by stress or by interference from other plants, RA is often less severely affected. Groundsel grown in a range of pot sizes responded by a sevenfold variation in total plant weight, but RA remained at about 21 per cent in all treatments (Harper and Ogden 1970). On the other hand in experiments with another annual plant, *Chamaesyce hirta*, and with an iteroparous perennial coltsfoot (*Tussilago farfara*), RA did vary significantly with plant density (Snell and Burch 1975; Ogden 1974). The addition of nutrients to the density treatments in the experiments with *Chamaesyce* ameliorated the reduction in RA caused by crowding.

Though it is conventional to measure RA in terms of the plant biomass allocated to reproductive and to non-reproductive structures, photosynthetically fixed carbon may not be the best unit in which to assess how a plant allocates available resources. If a mineral nutrient, rather than energy, limits plant growth, it would be more meaningful to measure how that nutrient is distributed between reproductive and vegetative organs of the plant. The fact that a nutrient application to dense populations of *Chamaesyce* can partially reverse the effects of density on RA strongly suggests that energy is not the only factor limiting how many seeds these plants produce.

Biomass and phosphorus (P) allocation to different organs of a monocarpic perennial ('biennial') alexanders (*Smyrnium olusatrum*) was measured in low-nutrient and in control conditions by Lovett Doust (1980) (Fig. 4.6). In the control treatment P was concentrated into reproductive organs quite disproportionately to their weight. However, the allocations of P and biomass to reproductive structures were quite

Fig. 4.5 Allocation of dry weight to the component parts of plant structure in *Senecio vulgaris* through the period from seedling to fruit production. (From Harper and Odgen 1970)

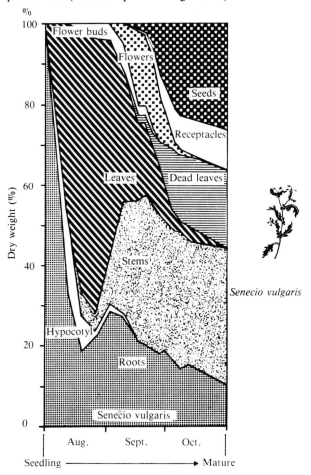

similar in the low-nutrient treatment. This suggests that P was not the limiting nutrient in impoverished soil. Selective addition of different nutrients to a low-nutrient treatment might have revealed which of the missing nutrients limited RA. Unfortunately this refinement was omitted from this experiment.

The general relationship between RA and the life history of species is shown in Fig. 4.7. Semelparous species generally allocate from 20 per cent to 40 per cent of their net assimilated energy to reproductive structures, while iteroparous species allocate from 0 to 20 per cent annually. This pattern may also be found within groups of closely

Fig. 4.6 The allocation of dry matter (a) and phosphorus (b) in control and low-nutrient treatments applied to plants of *Smyrnium olusatrum*. (From Lovett-Doust 1980)

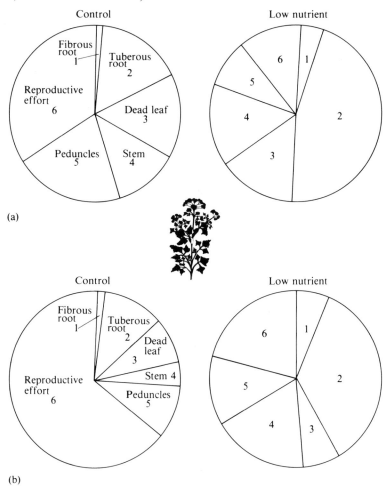

related species. Among the plantains (*Plantago* spp.) of North America, semelparous species produce a greater weight of seeds per unit area of leaf than iteroparous species (Primack 1979).

Cultivated plants grown for their grain have the greatest RAs of all those which have been measured. Most of these crops have been selectively bred from wild annuals which allocate lower proportions of biomass to reproduction and more to vegetative structures. This suggests that natural selection has not operated to maximize RA in the wild progenitors of crops, although it is axiomatic that it has selected those genotypes which leave the most descendants. This leads us to the conclusion that, in the wild, fitness may not always be maximized by maximizing RA. This is not the paradox that it may at first appear, because successful reproduction in natural populations often depends upon a plant competing successfully with its neighbours before it can produce any seeds. In dense, natural populations, it may therefore be advantageous to divert extra resources into roots and leaves, whereas

Fig. 4.7 The proportions of annual net assimilation involved in allocation to reproduction in different groups of flowering plants. (From Harper 1977, after J. Ogden)

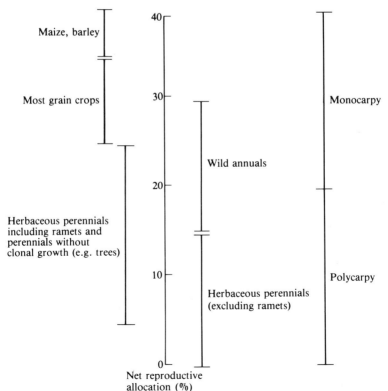

this may not be the case in agricultural systems where plant density is controlled.

Plant breeders have produced 'leafless' varieties of peas (which still have photosynthetically active stems and stipules) which make it possible to estimate the cost of leaf production in terms of lost seed yield per plant. Leaflessness is controlled at a single genetic locus. Conventional and leafless peas which were otherwise nearly genetically identical were grown in pure stands at densities of 16, 25, 44, 100 and 400 plants m^{-2}. Seed production per plant was higher for conventional peas at densities of 16 and 25 plants m^{-2}, but at all higher densities the leafless plants outyielded the leafy variety (Snoad 1981). If leafless peas produce more seeds at high density than leafy ones, why has the leafless gene not spread through natural population of *Pisum*? A major reason must be that such a gene would only confer increased fitness on a plant if all its neighbours were also leafless. A leafless plant in a field of leafy peas and weeds would leave few progeny!

Reproductive value

An increase in RA is not necessarily equivalent to an increase in the number of seeds produced per plant because seed production involves variable resource costs according to the size and energy content of seeds and fruits, flowers and other reproductive structures. Furthermore, density-dependent mortality among siblings may proportionally reduce the effective number of offspring produced as RA increases. The actual number of offspring produced (b_x) may vary directly with RA (Fig. 4.8(a)), or it may increase more rapidly than RA (Fig. 4.8(b)) or less rapidly than RA (Fig. 4.8(c)).

The optimum reproductive effort a plant expends in a particular year will depend upon how RA affects the lifetime balance between reproduction and survival. An increase in RA carries with it an increased risk of death and the accompanying risk of losing further opportunities to reproduce if death occurs. Thus we can expect the cost of reproduction to be a factor which influences the frequency of reproduction, in addition to any of the effects of juvenile and adult mortality already discussed.

If the life history is optimal, the RA at each age is such that the sum of current reproduction and likely future offspring is at a maximum. Since fecundity changes with age, the value of the sum of these two components of fecundity obviously depends upon how old a particular individual is. This determines how many more seasons it is likely to survive and how many offspring it may produce in the lifetime it has remaining.

The relative contribution an average individual aged x will make to the next generation before it dies is its *reproductive value* V_x (Fisher

Fig. 4.8 (a)–(c) The number of seeds produced in year x (b_x) may increase in direct proportion to increases in reproduction allocation as in (a); or it may increase faster (b), or slower (c) than RA.

Fig. 4.8 (d)–(e) The number of seeds produced in year x (b_x) and the residual reproductive value ($l_{x+1}/l_x V_{x+1}$) in relation to RA. The total lifetime seed production is the sum $b_x + (l_{x+1}/l_x V_{x+1})$ and is shown by a dashed line.

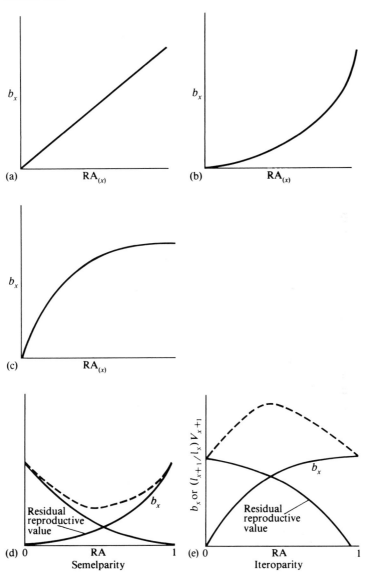

1930). In a stable population, V_x may be calculated as:

$$V_x = b_x + \sum_{i=1}^{i=\alpha} (l_{x+i}/l_x)b_{x+i}$$

where l_x is the survivorship to age x and b_x is the average fecundity of plants aged x. These statistics are taken from the life table and fecundity schedule of a population. In words: V_x is the sum of the average number of offspring produced in the current age interval (b_x), plus the average number produced in later age intervals (b_{x+i}), allowing for the probability that an individual now of age x will survive to each of those intervals (l_{x+i}/l_x). A graph of reproductive value against age for the population of *Phlox drummondii* described in Chapter 2 (p. 15) is shown in Fig. 4.9.

The importance of the later reproduction component depends upon the age of the plant and its *residual reproductive value*. After a plant has reproduced for the first time, its reproductive value might usually be expected to fall as it approaches the end of its lifespan. The *residual* reproductive value of a plant which reproduces in season x is equivalent to the chances which remain to it to produce further offspring in following seasons. In other words it is the probability of living one more season (l_{x+1}/l_x), times the reproductive value of a plant one season older ($x + 1$ seasons old) which we can denote by V_{x+1}. Therefore: Residual reproductive value $= (l_{x+1}/l_x)V_{x+1}$.

While we can generally expect b_x to increase with RA, V_{x+1} will decrease because the antithesis of reproduction and growth results in a strong association between high reproductive effort and short life (Fig. 4.7, p. 85). The same effects of reproduction on growth and survival will also reduce the probability of a plant living an additional season as RA increases. We can calculate the total number of offspring

Fig. 4.9 Reproductive values V_x for *Phlox drummondii*. (From Leverich and Levin 1979)

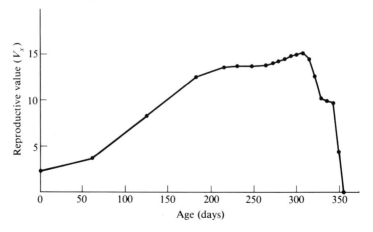

produced for a given RA by plotting the relationship between b_x and RA and $(l_{x+1}/l_x)V_{x+1}$ and RA on the same graph (Fig. 4.8(d), (e), p. 87. The most number of offspring are produced for the value of RA which corresponds to the largest value of the sum $b_x + (l_{x+1}/l_x)V_{x+1}$.

We will assume that natural selection favours the life history which produces the largest total number of offspring in the whole lifespan of a plant. When the curves of b_x and $(l_{x+1}/l_x)V_{x+1}$ on this graph are concave (Fig. 4.8(d), p. 87), a maximum number of offspring is produced when RA = 0 or 1. In other words, if this plant is to reproduce at all (i.e. if RA is to be greater than zero), it is best for it to commit all its resources to reproduction at once instead of spreading them out. This is the semelparous life history. The iteroparous life history produces most offspring when the b_x and $(l_{x+1}/l_x)V_{x+1}$ curves are both convex (Fig. 4.8(e) p. 87).

Another way of looking at the effect of the relationship between current reproduction and residual reproductive value on the evolution of semelparity and iteroparity is to plot these two values against each other. The exact form of the inverse relationship between reproduction now and reproduction later should be different for iteroparous and semelparous populations if our reasoning so far is correct. A concave relationship is to be expected for semelparous plants (Fig. 4.10(a)) and a convex curve for iteroparous ones (Fig. 4.10(b)).

What are the ecological conditions which will result in curves of these shapes? Generally speaking, any circumstances which reduce the resource cost per seed (number of seeds per unit of RA) as the size of a seed crop increases, or which increase the probability that a seed will itself survive to reproduce as the size of the seed or seedling cohort increases, will produce a curve of the type in Fig. 4.10(a). The way in which seed predators can exercise this kind of effect has already been mentioned. Density-dependent seedling mortality between seeds from

Fig. 4.10 Two forms of the relationship between current fecundity (b_x) and residual reproductive value.

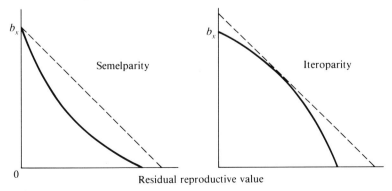

the same mother will produce the opposite effect and result in a convex curve of b_x versus $(l_{x+1}/l_x)V_{x+1}$.

The genus *Agave* contains a number of semelparous perennial species which grow in the deserts and chaparral of western North America. Some *Agave* species produce vegetative bulbils and consequently the genet is to be considered iteroparous. On the other hand, individual rosettes of most of these plants are semelparous. A typical species is the century plant (*A. deserti*) which delays reproduction for many years, storing water and carbohydrates in its rosette leaves. When it finally flowers, a rosette of only 60 cm in diameter is capable of producing an inflorescence up to 4 m tall. This feat can only be achieved by a massive translocation of water and assimilates from the rosette into the growing inflorescence which obtains 60 per cent of its biomass from this source.

Agaves have a similar habitat and morphology to plants in the genus *Yucca* but rosettes of these species are mostly iteroparous. Rosettes of yuccas and agaves both have stiff xeromorphic leaves and bear a central spike of insect-pollinated flowers. In an attempt to explain the difference in reproductive habit between species in these two genera, Schaffer and Schaffer (1977, 1979) compared the life history of seven *Agave* and five *Yucca* species. A sample of individual rosettes from each species was marked and the half-life of these sample populations was determined for the period after flowering. One *Yucca* (*Y. whipplei*) proved to have semelparous rosettes and one of the agaves had iteroparous ones (*A. parviflora*) Table 4.1. This demonstrated that differences in reproductive habit between the two genera were not

Table 4.1 Post-flowering half-life (months) and slope of the regression line of percentage of flowers developing into fruits, on inflorescence height (M_f) for twelve species of *Agave* and *Yucca*.

Agave			Yucca		
Species	Post-flowering half-life (months)	M_f	Species	Post-flowering half-life (months)	M_f
SEMELPAROUS			SEMELPAROUS		
A. utahensis	1.5	0.24	Y. whipplei	3.7	0.08
A. deserti	2.0	0.15			
A. chrysantha	2.5	0.18	ITEROPAROUS		
A. toumeyana	3.0	0.20	Y. glauca	48	0.00
A. palmeri	5.4	0.31	Y. utahensis	56	−0.02
A. schottii	7.8	0.17	Y. elata	70	0.01
			Y. standleyi	—	0.00
ITEROPAROUS					
A. parviflora	29.7	−0.03			

From Schaffer and Schaffer 1977

simply a result of different evolutionary descent, but had evolved separately in each genus.

Schaffer and Schaffer suggested that semelparity in these plants was favoured by the selective behaviour of pollinating insects which they showed visited the largest inflorescences disproportionately more often than smaller ones. They calculated the pollination advantage gained by plants with large inflorescences in different species, by plotting the number of pollinator visits observed per centimetre of flower stalk against inflorescence height. The slope (M_p) of such a graph is steep when the frequency of pollinator visits accelerates with increasing height and shallow if it does not accelerate. They then showed that the proportion of flowers which produce ripe fruit per centimetre of stalk was also related to inflorescence height and that the slope of this relationship (M_f) was correlated with M_p.

Assuming, as this evidence suggests, that pollination was the main factor determining the proportion of flowers which produced fruit, M_f was then used as a measure of pollinator preference for larger inflorescences. The relationship between M_f and post-flowering half-life for eleven species of *Yucca* and *Agave* showed that semelparous species had the highest values of M_f (Fig. 4.11). Hence these species acquired a greater benefit from increasing inflorescence height because the larger the inflorescence, the more seeds per centimetre were produced.

It may be concluded that this was probably a major evolutionary factor causing them to concentrate their reproduction in a single large

Fig. 4.11 Slope, M_f, of the regression line: percentage of flowers developing into fruits versus stalk height, plotted against post-flowering half-life, PFHL, for eleven species of yuccas and agaves $r = 0.77$; $p(r) < 0.005$, one-tailed. (From Schaffer and Schaffer 1977)

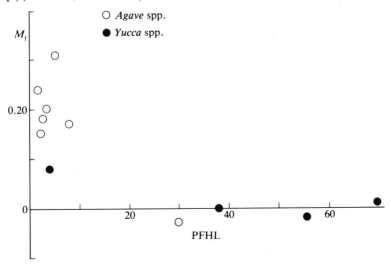

inflorescence, reducing the residual reproductive value of a rosette after first flowering to zero. The iteroparous rosette species, including the iteroparous *Agave*, gained no increase in seed set per centimetre of inflorescence in larger stalks compared with smaller ones. Hence the selection pressure in favour of one large burst of reproduction was absent from these populations.

Seed size, clutch size and crop size

We have so far examined, and attempted to explain in evolutionary terms, the variation to be found in the timing of reproduction and the effort expended in producing seeds. This reproductive expenditure is packaged into seeds of varying size, seeds are packaged into reproductive units or clutches (e.g. the fruit of angiosperms) containing different seed numbers and both packages combine to produce seed crops of different total sizes. Can variations in seed size, the number of seeds per clutch and the size of seed crops be explained as adaptations which increase the number of descendants a plant leaves?

An enormous range in seed size occurs between species, from the seeds weighing 10^{-6} g produced by the orchid *Goodyera repens* to the seed of the double coconut (*Lodoicea maldivica*) which weighs in at over 10^4 g (18–27 kg) (Harper, Lovell and Moore 1970). A number of comparative studies of seed size have been made and these suggest that this character has been adjusted by natural selection in various ways, depending upon the life history and habitat of species. The mean weight of seeds increases progressively through herb, shrub and tree species in the flora of the British Isles and that of California. It also follows the same trend on a world-wide scale (Table 4.2).

Within the herb group in California, the mean seed weight of annuals is significantly less than that of perennials (Baker 1972). This difference does not occur in Britain (Salisbury 1942; Hart 1977) when the floras of all habitats (woodland, grassland, etc.) are lumped together. Trends of seed size with variations in life history are likely to be obscured when a whole flora is compared because of the stronger association of seed size with habitat.

Table 4.2. Seed weight in relation to plant habit.

Mean seed weight (g)			
Habit	Britain	California	World-wide
Herbs	0.0020	0.0057	0.0070
Shrubs	0.0854	0.0075	0.0691
Trees	0.6534	0.0096	0.3279

From Levin and Kerster 1974

A comparison of seed size for annuals and perennials which occur in the same habitat, enables habitat to be eliminated as a variable. When such a comparison is made for plants in the flora of British calcareous grasslands for instance, annuals are found to have significantly smaller seeds then perennials herbs (Silvertown 1981b). The annuals in this grassland sample were semelparous while most of the perennials were

Fig. 4.12 Fruits of two annual species of *Galium* drawn to the same scale. Above, the fruits of *G. anglicum*, a species characteristic of open situations on sandy soils, especially in East Anglia. Below a single fruit of the goosegrass (*G. aparine*), a species of woodland margins, scrub, and hedgerows. The fruits of the scrub species weigh nearly 250 times those of the open habitat species. (From Salisbury 1942)

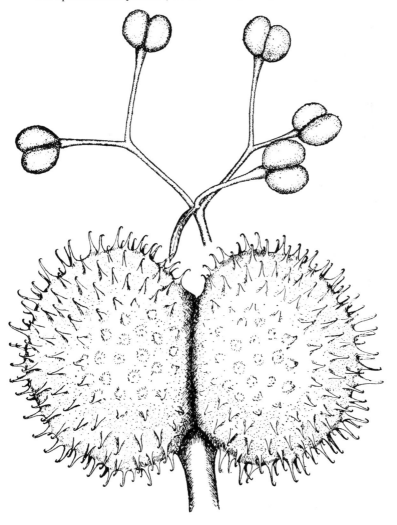

iteroparous. It seems likely that natural selection has increased seed number at the expense of seed size in these annuals and that this shift compensates for the disadvantage of semelparity which we discussed on p. 82.

Herbs of woodland and scrub in Britain generally have larger seeds than herb species in the same genus which grow in more open habitats such as grassland or arable fields (Salisbury 1942). Compare for instance the seed size of *Galium anglicum*, an annual plant of open habitats in East Anglia, with that of *G. aparine* which grows in more shaded habitats and which is also annual (Fig. 4.12). This difference is presumably the result of the increased food reserves required for seedling establishment in shade, a requirement which is reflected in higher seedling mortality among species with smaller seeds in experimental shade conditions (Fig. 4.13).

Variations in seed weight between species in the Californian flora appear to be more strongly related to the risk of seedling mortality due to drought than due to shade, and a positive relationship between seed weight and the dryness of the habitat occurs both among herb species within the same genus and for whole herb communities. A similar relationship between moisture availability and seed weight is also found in Californian trees. The explanation offered for these relationships is that a larger seed enables a seedling to produce a more extensive root system and thus obtain water more rapidly and efficiently than a small seed in a dry environment. California has a large adventitious flora of

Fig. 4.13 The relation between death-rate in shade conditions and log mean seed weight in nine tree species. (From Grime and Jeffrey 1965)

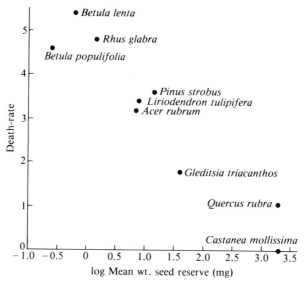

introduced species which fit the same patterns of seed weight and
environmental conditions shown by native species (Baker 1972).

This correspondence between the behaviour of adventitious and
native species has some interesting and far-reaching implications for
how we view the adaptive fit between seed size and environment (e.g.
species with large seeds occurring in shaded habitats). Two hypotheses
for this fit are possible: 1. species have evolved a seed size characteristic
of a particular habitat largely under selective forces operating within
that habitat; or 2. species with seeds ill-suited to regeneration in a
particular habitat suffer ecological displacement. In other words, species
which happen to have large seeds become woodland plants when they
invade an area. Invading species with small seeds cannot occupy
woodland but can become weeds or grassland plants. The process of
ecological displacement must explain the distribution of most of the
adventitious plants of California. For how many native species could
this also apply?

In sharp contrast to the large variation in seed size which occurs
between species, mean seed weight within species tends to be highly constant and the plant character least affected by density (see Ch. 5 p. 129).
So invariable is the seed weight of the carob tree (*Ceratonia siliqua*)
that its seeds have been used as units of weight (the carat) for trading in
gold. The reason plants generally respond to density by reducing seed
number rather than seed weight is easy to see. Small reductions in the
energy used to provision an individual seed would place a seedling at a
severe competitive disadvantage. This would produce a reduction in the
parent's fitness which could not be outweighed by the production of
additional seeds, each with the same handicap. The consequences of
small seed size in a competitive situation are demonstrated by an

Fig. 4.14 The percentage of the leaf area (a) and the percentage of the
light interception (b) of plants of *Trifolium subterraneum* from large and
small seeds grown in a mixed sward. (From Black 1958)

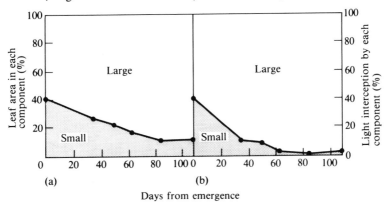

Days from emergence

experiment in which large and small seeds of subterranean clover (*Trifolium subterraneum*) were planted in a mixture (Fig. 4.14). Mortality selectively eliminated seedlings from small seeds, reducing their percentage share of light interception to virtually zero after 80 days.

The principle that a few propagules of normal size have a better prospect of survival than the combined prospects of a lot of 'runts' applies in the animal kingdom too where it has been developed into a theory of clutch size, principally employing data on birds (Lack 1954; Cody 1966). One prediction of this theory which concerns us here is that when the length of the growing season is reduced with geographical latitude or altitude, plants should produce fewer seeds (smaller clutches) rather than smaller ones. Johnson and Cook (1968) measured clutch size in *Ranunculus flammula*, a perennial growing at a range of altitudes from near sea-level to 1700 m in the Cascade mountains of western USA. The number of carpels per ramet was counted in populations at seven altitudes, each with a progressively shorter growing season, and populations obtained from five altitude stations were planted in experimental conditions. Plants from higher altitudes had fewer carpels per ramet than lower populations and maintained these differences when grown in uniform experimental conditions, supporting the hypothesis that clutch size was genetically adapted to the shorter growing season (Fig. 4.15). Johnson and Cook omit to mention the weight of carpels from the population they studied or the number of ramets per plant, without which information we cannot be sure that *R. flammula* behaves fully as the clutch theory would predict.

Fig. 4.15 Mean seed production per ramet for *Ranunculus flammula* collected at five elevations and mean seed production per genet for *Heloniopsis orientalis* at five altitude stations. (Data from Johnson and Cook 1968, and Kawano and Masuda 1980, respectively)

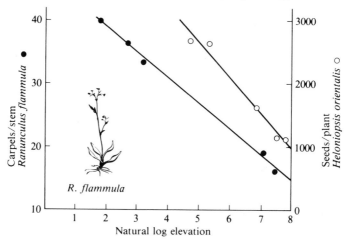

A study of reproduction in five populations of an evergreen perennial lily (*Heloniopsis orientalis*) growing at elevations from 100 m to 2600 m was made by Kawano and Masuda (1980) in Japan. They found a trend of decreasing seed production with the length of the growing season and with altitude in *Heloniopsis*, similar to that in *R. flammula* (Fig. 4.15). However, Kawano and Masuda's measurements were made on populations *in situ* and not under controlled or garden conditions. It is therefore possible that their measurements reflect a purely phenotypic response to altitude and not genetic adjustments of clutch size.

Kawano and Masuda also measured RA in these populations. Reproductive allocation increased from 10 per cent at 100 m elevation to nearly 25 per cent at 2600 m, despite the fact that fewer seeds were produced at the greater elevations. In fact, the energetic cost of producing a seed (seed number per unit of RA) increased sixfold between 100 m and 2600 m. This was because the ancillary reproductive structures, the flower stalk and its attached leaf, the capsules, the petals and the sepals, received a relatively constant amount of biomass at all elevations. These organs apparently represent the fixed energetic costs of seed production in this species. It is non-reproductive organs and seed number, in that order, which are the components of plant structure that are adjusted when resources are in short supply. The weight of non-reproductive organs must, to some extent, be related to the chances of a plant surviving and flowering in the next year (i.e. residual reproductive value).

An interesting evolutionary conundrum exists in the way in which milkweeds in the genus *Asclepias* package their reproductive output. These plants indulge in a prodigal production of flowers, some species typically carrying over 100 per plant, of which only a fraction (around 1%) ever produce seeds, even when all are fertilized (Wyatt 1976). This extraordinarily low ratio of seed production to flower production would seem to be a 'waste' of reproductive effort. Could it be an unavoidable consequence of the way in which plants alter their reproductive output under selection pressures?

The reproductive parts of *Asclepias* can be divided into the following hierarchy of components (Fig. 4.16): the weight of individual seeds; the number of seeds in a pod; the number of ripe pods per umbel of flowers produced; the number of umbels per stem; and the number of stems per individual plant. Wilbur (1976, 1977) compared the relative size of each of these components in seven species of milkweed which grew together in an experimental reserve in Michigan and showed that there were statistically significant differences within the genus in the way in which a given quantity of seed production was packaged. Unfortunately RA as defined earlier in this chapter was not measured in this study so we will use the term *reproductive output* to refer to the absolute quantity of seeds produced.

Asclepias tuberosa, A. verticillata, A. purpurescens and A. exaltata produced similar numbers of seeds per stem but these were packaged in large numbers in a few pods in A. exaltata and in smaller numbers in more pods in the other three species (Fig. 4.16). If we compare A. exaltata and A. purpurescens which produce similar numbers of seeds per plant we find that the total weight of seeds per pod is also about the same in these two species (436 and 425 mg respectively), but that this is made up of small seeds in A. purpurescens and larger ones in A. exaltata. Does this mean that seed size and seed number per pod evolved as alternative forms of reproductive packaging in Asclepias? A glance at the pattern of reproductive packaging in A. verticillata (Fig. 4.16), the species with the smallest seeds (2.14 mg) of all *and* the smallest number per pod (42), removes this idea. Indeed it is very difficult to see any clear pattern at all which could explain the differences in reproductive output and reproductive packaging in this group of species.

Fig. 4.16 The packaging of reproductive output in seven species of *Asclepias*. (Data from Wilbur 1976)

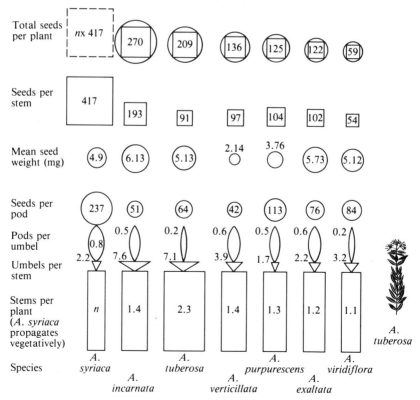

The plasticity of plant growth provides plants with more options and more flexibility in altering reproductive output to meet prevailing circumstances than are open to most animal species. This plasticity is sometimes itself interpreted as a response to natural selection (Bradshaw 1965). Plasticity can be expressed in different ways according to how reproductive allocation or reproductive output is adjusted in relation to energy constraints. Two different types of plastic response were shown in the asclepiads studied by Wilbur (1977). Three species, *A. incarnata, A. tuberosa* and *A. exaltata*, varied total reproductive output by altering the number of seeds produced through changing the number of pods aborted per umbel. However, *A. verticillata* and *A. viridiflora* varied total reproductive output by breaking the golden rule of constant seed weight and altered the mean weight per seed instead of any of the other components of reproductive output, which would have produced changes in seed number. Though several different factors including predation, interference from other plants and variations in the predictability of the habitat were studied (Wilbur 1976) in an attempt to explain all these differences in *Asclepias* life history, no satisfactory 'adaptive' explanations have emerged to account for most of them.

Wilbur attempted to do what most ecologists interested in evolution try to do – to explain different patterns of life history as adaptations evolved under different selection pressures. Though this is often possible, there may be a good reason why the differences in reproductive packaging in *Asclepias* cannot be accounted for in this way. Instead of being different, appropriate ways of adapting to different selection pressures, the variety of reproductive packaging in *Asclepias* may simply reflect a species' own unique and different ways of adapting under *the same* selection pressure. In other words we must recognize that one 'problem' may have more than one 'answer'. What is the 'problem' to which *Asclepias* species have devised so many responses?

A clue lies in the characteristic of reproductive packaging that all the species share – the massive abortion of fruits. Why develop so many flowers, only to throw most seed pods away after they have been fertilized? Two answers have been suggested. The first is that a large inflorescence is necessary to attract pollinators. The second is that flowers which are pollinated but which abort are actually acting as 'pollen donors' (i.e. males) and that their female function is redundant. These hypotheses suppose that, in effect, most of the flowers on an *Asclepias* umbel are not functional seed-producing organs, but flags put out to attract pollinators and/or they are functional males. In both cases it is assumed that plants are physiologically incapable of sustaining the development of all pods and that therefore they abort most of them, once flowers have attracted pollinators and released pollen.

If we take either or both of these explanations for the production of 'excess' flowers as correct, or even if we do not but we still recognize that some fruit must be aborted, it no longer seems very surprising that when plants abort the offspring they cannot support, different species do this in different ways. Fruit and ovule abortion is quite widespread in other species too, and is achieved by various routes. It remains to be seen how far one means of abortion increases fitness more than another in specific ecological conditions (Lloyd 1981).

Finally, what evidence is there that either hypothesis to explain the production of large inflorescences is correct? The crucial experiment to test the pollinator attraction hypothesis is to remove some flowers from umbels and then to measure pollinator visits and seed set per remaining flower. This experiment was done on *A. syriaca* by Willson and Rathcke (1974). They found that inflorescences with twenty flowers produced the greatest number of pods per flower but that smaller and larger inflorescences were less efficient producers of fertilized pods. A tenfold increase in flower number (20 to >200 flowers per inflorescence) produced only a fourfold increase in the number of pods. These results do not suggest that the commonest flower number per inflorescence found in Willson and Rathcke's population was that which maximized fecundity. Wyatt (1980) has also found an inverse relationship between fruit set per flower and the numbers of flowers in an inflorescence in *A. tuberosa*.

Many other species with large inflorescences also abort substantial fractions of their flowers. The appropriate experiment to test the pollinator attraction hypothesis has been done on one of these, the western catalpa *Catalpa speciosa*. Inflorescences of this tree carry an average of twenty-seven flowers, but only about three produce fruit, even when all have been pollinated. In experiments, smaller inflorescences from which flowers had been removed, produced a lower ratio of fruit to flowers than larger ones (Stephenson 1979).

The evidence that *Asclepias* flowers disperse pollen before they abort is good (Willson and Price 1977; Willson and Bertin 1979). The tale of reproductive packaging in asclepiads is a cautionary one for the ecologist who seeks an adaptive explanation for every difference between species. Differences between related species may result from chance historical events or from morphological constraints and are not always functional (Gould and Lewontin 1979).

Differences in the way in which seeds are packaged may or may not reflect selection pressures exerted at some time in the past. Total crop size is also a variable of plant fecundity which often changes from year to year within populations. The magnitude of this variation can be very different in different species. In particular, some tree species (e.g. *Fagus* spp., *Quercus* spp., *Pinus* spp.) produce vast crops of seed (mast) in some years, but few seeds in the intervening periods between mast years. Typically, seed production by different individuals in such populations is synchronized and mast years are correlated with climatic variables. In Europe, for instance, the beech (*F. sylvatica*) may mast in the year following a hot summer but rarely in a year following a cold one.

Two hypotheses have been advanced to explain masting behaviour. The first is simply that climatic conditions suit seed production better in some years than in others and that barren periods result because trees take time to recover from the effort of reproduction (Fig. 4.1, p. 74). The second hypothesis, prompted by the observation that masting seems to waste opportunities for reproduction, is that these lost opportunities are in some way compensated because the habit increases the fitness of individual trees which vary their seed production in this way.

The argument is that seed predators consume a large proportion of small seed crops but that they cannot consume a tree's entire crop in a mast year. Hence the probability of a seed escaping predation is greatest when crops are large. However, it would be disastrous for the tree if large crops were produced regularly because predators would simply build up their numbers from one year to the next on succeeding bumper crops. This hypothesis predicts that there should be a negative relationship between the probability of a seed being eaten and the size of the current seed crop in a masting species. This prediction is confirmed by information collected by foresters on a number of species (Silvertown 1980b). The effect of two predators on seed survival in ponderosa pine is shown in Fig. 4.17. Unfortunately, however, the argument about the adaptiveness of masting does not stop here because of the correlation between seed crop size and various climatic factors.

The problem with the climatic explanation is that it does not explain why individuals in some species exhibit the masting habit more intensely than others in the same geographical region. Therefore the adaptive hypothesis might be strengthened if it can be shown that the masting habit is most pronounced in those tree populations where seed predation is strongest.

Using data on variation in the annual seed production of trees and data on seed predation in crops of different sizes (e.g. Fig. 4.17) for a range of species, it is possible to test the idea that masting species are attacked more severely than non-masting species when they produce small seed crops. Comparing seed production and seed predation data for fifteen species in this way, Silvertown (1980b) found that five of the seven most heavily preyed-upon species showed the masting habit. Among the eight species which suffered lower seed predation, only two showed very variable seed production.

Another intriguing piece of evidence which supports the idea that masting is a defensive strategy which protects trees from seed predation, comes from a comparison of the seeding behaviour of two populations of the tropical tree *Hymenaea coubaril*. This tree occurs both on the island of Puerto Rico where one of its major insect seed predators is absent, and on mainland Costa Rica where these predators are present.

The mainland tree population shows the masting habit but the island one does not. Other morphological features of *H. coubaril* fruit which help deter predators in Costa Rica are also absent from populations in Puerto Rico (Janzen 1975b).

It follows from the argument that masting prevents animals consuming an entire tree crop, that trees with fleshy fruits and animal dispersed seeds should not mast but should produce fruit regularly. The seeds in

Fig. 4.17 The relationship between annual cone crop size and the probability of a seed of ponderosa pine escaping predation by (a) chalcid wasps, or (b) abert squirrels. Study (a) was done in California by Fowells and Schubert (1956), study (b) in Arizona by Larson and Schubert (1970).

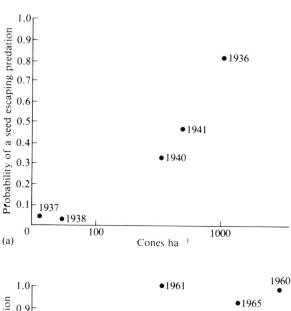

(a)

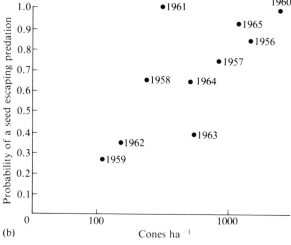

(b)

such fruits generally pass through the gut of the dispersal agent intact, so that the only effect of masting in a fleshy-fruited species would be to prevent seed dispersal. The hypothesis that fleshy-fruited species do not mast may be tested by using a method of between-species comparison as before. A comparison of this kind, using species from the North American sylva, confirms the hypothesis and shows that most trees with non-fleshy dispersal units mast to some degree, while most of those with fleshy dispersal units do not (Silvertown 1980b).

Patterns of reproduction, dispersal and persistence

The life history variables we have examined in this chapter are found in almost all combinations. There are both iteroparous and semelparous perennials for instance, but certain combinations of life history characteristics appear to occur more often and others less often than we would expect by chance. The comparative rarity of semelparous perennials, for example (Hart 1977), indicates that conditions which favour the evolution of prolonged life generally also favour iteroparity. The apparent tendency for plants and animals to express sets of correlated life history characters has led many authors to describe these sets as *strategies*.

The most influential of these authors, MacArthur and Wilson (1967), distinguished two contrasting types of life history strategy based upon those which are adapted for dispersal and those adapted for persistence. They named organisms adapted for colonization and reproduction in expanding populations *r*-strategists and those adapted for persistence and reproduction in stable populations *K*-strategists. These terms are taken from the logistic model of population growth in which *r* is the intrinsic rate of natural increase (or the instantaneous rate of population growth, unlimited by density) and *K* the carrying capacity (as defined in Ch. 1):

$$\frac{\mathrm{d}n}{\mathrm{d}t} = rN \frac{(K - N)}{K}$$

Gadgil and Solbrig (1972) envisaged the two strategies as opposite ends of a continuum with *r*-strategists reproducing faster than *K*-strategists in expanding populations at low density, and competition leading to the reverse outcome in stable, high-density populations. This idea is summarized in Fig. 4.18.

In a population of plants composed of genotypes expressing the *r*-strategy and others behaving as *K*-strategists, competition should lead to *r*-selection in unstable environments and *K*-selection in stable ones. A number of authors (e.g. Hairston, Tinkle and Wilbur 1970; Stearns 1977) have pointed out that *r* and *K* are not simple life history parameters (such as RA) but are determined by the interaction of a number of factors. The intrinsic rate of natural increase (*r*) cannot be increased directly by natural selection, only selection on its components – the instantaneous birth-rate (b) and instantaneous death-rate (d) – can do this. The carrying capacity (K) is similarly not a straightforward life history parameter but the result of an amalgamation of environmental and demographic factors. Valid though these objections (and several others) are to the simplistic view that natural selection should maximize r or K (but not both) in a population, we may still examine the *hypothesis* that some organisms are adapted for dispersal, others for persistence and that no plants can be successfully adapted for both extremes of environmental stability. Though it may strictly be erroneous to call these two strategies r and K, we will follow convention and do so to avoid confusion.

The predictions of the r- and K-model are that plants of stable environments (K-strategists) should possess or evolve slow development, postponed reproduction, iteroparity, small RA, small crops of large seeds and long life. In contrast, r-strategists should show rapid development and early reproduction, semelparity, large RA, large crops of small seeds and short life. These predictions are listed in part (a) of Table 4.3. It is an assumption of r- and K-theory which is not always stated clearly that environmental instability affects adult rather than juvenile survival. In stable environments the theory predicts that there will be heavy density-dependent mortality among juveniles and that this will consequently select for large seed size (see p. 95). The theory also makes some predictions which contradict the life history theory we discussed earlier in this chapter. This suggested that high mortality should favour early reproduction. On the other hand the

Fig. 4.18 The population growth-rate as a function of population density for two biotypes; (r) = growth-rate of the population unlimited by density; (K) = carrying capacity; (d) = population density at which population growth of the two biotypes are equal in their value. (From Gadgil and Solbrig 1972)

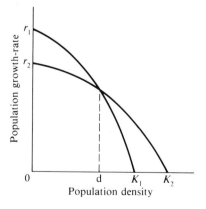

mortality which accompanies *K*-selection is supposed to favour delayed maturity.

The assumption that environmental fluctuations affect adult rather than juvenile survival is crucial to the *r*- and *K*-hypothesis whose predictions are turned on their head if we assume that such fluctuations would affect seed production or seedling survival more than adult survival (Murphy 1968; Schaffer 1974). Under these circumstances, plants should adopt the strategy predicted by the theory of 'bet-hedging' and should spread reproduction over several seasons (hence the name of the theory) and adopt the other characteristics of low juvenile survival shown in part (b) of Table 4.3. The theory of 'bet-hedging' appears to fit the life history patterns of some marine fish such as the Pacific sardine (Murphy 1968), but has scarely been considered or tested as an explanation of plant life history strategies. On the other hand the *r*- and *K* theory has attracted a lot of attention from plant population ecologists.

The most thorough test of the *r*- and *K*-theory in plants (Gadgil and Solbrig 1972; Solbrig and Simpson 1974, 1977) examined the life history of the common dandelion *Taraxacum officinale* growing in three populations subject to varying degrees of environmental disturbance and density-independent mortality. The dandelion populations studied were apomictic and hence were composed of a number of distinct clones which could be identified by electrophoresis. Of the four biotypes (A, B, C and D) identified by this technique, three (A, B, C) occurred in all three populations and the fourth (D) was absent only from the most

Table 4.3. The life history characters predicted for plants by the *r*- and *K*-selection thoery and by 'bet-hedging'.

Life history character	Stable environments	Unstable environments
(a) *r*- and *K*-selection – adult mortality variable		
Development	Slow	Fast
Age of first reproduction	Late	Early
Reproductive allocation	Small	Large
Number of seed crops	>1 (Iteroparity)	1 (Semelparity)
Size of seed crops	Small	Large
Size of seeds	Large	Small
Adult longevity	Long	Short
(b) 'Bet-hedging' juvenile mortality variable		
Development	Fast	Slow
Age of first reproduction	Early	Late
Reproductive allocation	Large	Small
Number of seed crops	>1 (Iteroparity)	>1 (Iteroparity)
Size of seed crops	Large	Small
Size of seeds	Large	Small
Adult longevity	Short	Long

Modified from Stearns (1976)

disturbed site (Fig. 4.19(a)). Biotype A typified the *r*-strategist, being most abundant in the site of high disturbance (a footpath) and least abundant in the most stable site (an old pasture). Plants of biotype A also had a high reproductive output (Fig. 4.19(b)) and a low rate of mortality when grown in pure culture (Fig. 4.19(c)). Biotype D typified the *K*-strategist and was most abundant in the population with low disturbance (see Fig. 4.19(a)–(d)). Plants of biotypes B and C fell somewhere between A and D and the extremes of the of the *r*- and *K*-continuum in their habitat distribution and the production of flowers.

Interference experiments based on the replacement series design (see p. 152) were set up using plants from biotypes A and D sown at two densities. Biotype D suppressed plants of biotype A in 50/50 mixture but was not itself affected by interference from the *r*-strategist at high or low density (Fig. 4.19(e), (f)).

Another, long-term, experiment was set up to test the outcome of competition between biotypes A and D in disturbed and undisturbed plots. Plants of the two biotypes were sown in garden plots in mixtures of equal proportions. In one treatment all plants were defoliated periodically to simulate environmental disturbance. Disturbance was created in another treatment by tilling the soil in the second and third years after sowing. Plants from each treatment were sampled, 4 years after the initial sowing, and the ratio of the two biotypes was determined. The results of this experiment are shown in Fig. 4.19(g). Biotype D virtually replaced biotype A in the undisturbed plots. The reverse displacement took place in the disturbed treatments (Solbrig and Simpson 1977).

An important feature of this test of the *r*- and *K*-hypothesis was that it showed that differences between biotypes persisted when they were grown in standard conditions. Studies of other species such as *Polygonum cascadense* (Hickam 1975) have shown that life history differences may arise from the plastic response of plants to different environmental conditions. This could mislead the casual observer into believing that *r*- and *K*-selection had operated in a population where no substantial genetic differences actually existed between so-called '*r*-strategists' and '*K*-strategists'. The interpretation of a number of studies (e.g. Abrahamson and Gadgil 1973) is hindered by the lack of any evidence that '*r*'- and '*K*'-strategists are actually different when grown under standard conditions. In a critical review of the evidence for *r*- and *K*-selection, Stearns (1977) arrived at the conclusion that more experimental work is required to settle the *r*- and *K*-issue in both plants and animals.

Summary

Individual plant *fitness* is determined by reproduction and survival. The lifetime sum of these two components may be expressed as $\Sigma l_x b_x$.

Fig. 4.19 An experimental test of *r*- and *K*-selection in dandelions. (Data from Solbrig and Simpson 1974, 1977)

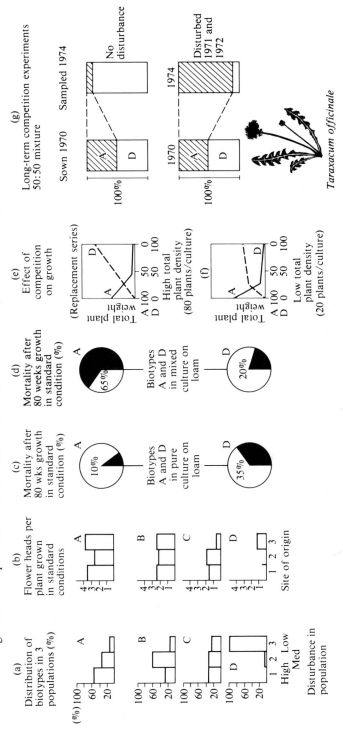

Reproduction and growth are alternative ways in which a plant may use limited resources. The relative allocation of these resources throughout the life cycle affects fitness.

Semelparity is favoured by a high ratio of juvenile/adult survival while *iteroparity* is favoured by a low ratio of these parameters. High mortality favours early reproduction.

Reproductive allocation describes the allocation of resources to sexual reproduction within 1 year.

Reproductive value is a measure of the number of successful offspring likely to be produced by an individual of given age before it dies.

Seed size, clutch size and crop size determine the fecundity of a female plant. Seed size is the least variable of these parameters. Crop size is the most variable. The evolutionary reasons for this may be found in the ecology of plants. The total fecundity of a plant is determined by a hierarchy of components which may be adjusted in different ways in different species, sometimes to achieve similar results.

Sets of correlated life history characters are known as *strategies*. It has been suggested that two important and opposing strategies maximize an individual's ability to disperse or to persist. This is the *r*- and *K*-hypothesis.

5
The regulation of plant populations

Plant populations are not dusty museums of plant life where the same faithful individuals are to be found on every visit, but show more the constant activity of a railway station; witnessing a never-ending flow of new arrivals and departures. The timetables of these plant arrivals and departures are fecundity schedules and life tables. In view of this constant flux of individuals through plant populations a remarkable observation emerges about the net outcome of all these changes: individuals come and go but population size often remains more or less constant.

Population regulation and density dependence

If the death-rates and birth-rates which determine a population's size were subject to random changes it would only be a matter of time before such a population became extinct. Random processes countenance extremes, oblivious of any catastrophic consequences. The first extreme increase in the death-rate caused by a grazing animal or by adverse climate would finish off a population for ever. This is no idle observation because it suggests, on theoretical grounds alone, that populations which have a high flux of individual members and which are stable in overall population size must be cushioned from the random occurrence of high mortality by the action of some specific process.

Populations may be cushioned from local catastrophe in two basic ways: 1. by the immigration of plants from areas not affected by the catastrophe (Ch. 8); or 2. by changes in fecundity or mortality in the population itself, which compensate for the effects of the catastrophe. The second of these is discussed in this chapter.

A clear example of a population showing overall stability of adult numbers from year to year despite the death of an entire generation of plants each year is seen in an 8-year study of a small, ephemeral, sand dune plant *Androsace septentrionalis* by Symonides (1979c) (Fig. 5.1). Each year during which she studied this plant in a population growing on inland dunes in Poland, Symonides found between 150 and 1000 seedlings m^{-2}. Each year mortality reduced the population by between 30 and 70 per cent. The remarkable thing is that this annual mortality never overshot a limit – at least fifty plants always survived to fruit.

Any populations, such as this one, which show the operation of clear limits on population size are said to be *regulated*. Population regulation operates via *density-dependent* processes which alter the numerical impact of fecundity or mortality on population size as population density changes. A density-dependent mortality factor is one that relaxes as population density declines, and thereby slows or halts population decrease. When population density increases, a density-dependent mortality factor kills an increasing proportion of the population. An example is seen in the relationship between seedling survival and the original density of seeds in the Wisconsin population of *Acer saccharum* studied by Hett (1971) (Fig. 5.2(a)). Density-dependent fecundity may also regulate population size by the production of fewer seeds per plant as population density rises (Fig. 5.2(b)).

The effect of all density-dependent processes is to produce populations of more or less constant density from populations which originally differ in size. Density-dependent mortality may reduce a wide range of seedling densities to a smaller range of adult densities, and density-dependent fecundity may result in similar numbers of seeds being produced by plant populations of widely different densities (Fig. 5.3(a)). The operation of both processes in some greenhouse populations of the grass *Bromus tectorum* planted at densities of 5, 50, 100 and 200 seeds per pot (Fig. 5.3(b)) illustrates how the combined effect of density-dependent mortality and density-dependent fecundity causes the population numbers to converge, so that approximately the same number of seeds begin the next generation, irrespective of their parent's planting densities (Palmblad 1968). Density-dependent mortality operates within a generation, reducing a large cohort of seedlings to a

Fig. 5.1 The population dynamics of *Androsace septentrionalis* over an 8-year period: (a) beginning of germination; (b) maximum germination; (c) end of seedling phase; (d) period of vegetative growth; (e) flowering; (f) fruiting. (From Symonides 1979b)

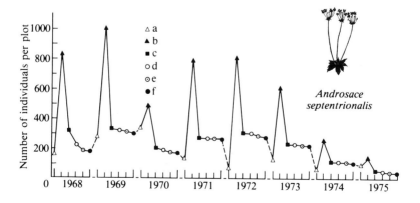

smaller cohort of adults. Density-dependent fecundity alters population size in a slightly less direct way.

Because density-dependent fecundity regulates the *seed* production of a population, it has an influence on the potential population size of the *next* generation of adults. In effect therefore, this is *delayed* density

Fig. 5.2 Density-dependent processes in two plant populations: (a) mortality in a population of sugar maple establishing from seed (Hett 1971); (b) fecundity in experimentally manipulated natural populations of *Vulpia fasiculata*. (Watkinson and Harper 1978)

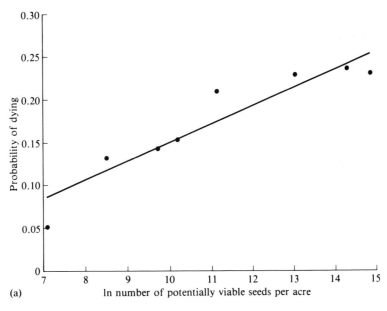

(a)

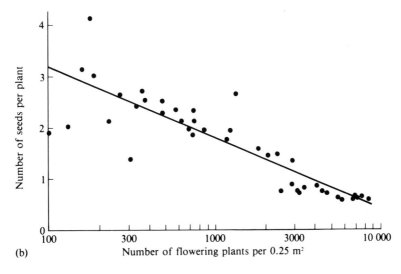

(b)

dependence: the effects of density upon mothers is visited upon their children.

Different density-dependent factors may regulate population size with varying effectiveness. Figure 5.4 illustrates a density-dependent mortality factor which *undercompensates* for population changes, one that *exactly compensates* for population changes and a factor that overcompensates. On a graph of mortality plotted against the density of the population on which it acts, these three types of density dependence are shown by slopes of <1, 1 and >1 respectively. Some mortality factors actually *decrease* as population density increases. Two examples of this *inverse density dependence* were seen in Chapter 4, where the proportion of *Pinus ponderosa* seeds eaten by both insects and squirrels decreased as the size of the seed crop increased. Inverse density-dependent factors do not regulate populations according to the definition of regulation we are using. Indeed, such factors have the reverse effect and may cause wild fluctuations in population size of the kind which can lead to extinction.

Fig. 5.3(a) An idealized diagram of density-dependent seedling mortality and density-dependent fecundity and their role in the regulation of a plant population. (b) An example of density-dependent mortality and density-dependent fecundity regulating population size in four experimental populations of an annual grass *Bromus tectorum*. (Data from Palmblad 1968)

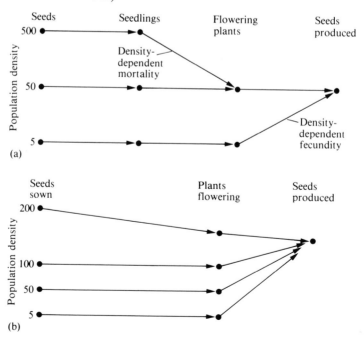

For convenience we will analyse mortality factors according to the stage of the life cycle, equivalent to a particular row in the life table, they affect. There are two general approaches to the problem of discovering which mortality factors are density-dependent regulators of particular plant populations. The first approach to the problem is experimental, and simply involves subtracting or adding individuals to one stage of the population and observing the effect of this on the number of individuals entering the next stage. For instance, Watkinson obtained the range of *Vulpia* population densities in Fig. 5.2(b), p. 111 by means of adding seeds to some populations and thinning others by removing seedlings. In this case varying densities had no effect on mortality but altered individual plant size with the result shown in the figure.

The second method of studying population regulation has been used a great deal for the analysis of insect population dynamics (Varley, Gradwell and Hassell 1973), but it requires an entire life table and fecundity schedule to be compiled for a discrete generation each year for at least 5 or 6 years. The lack of such data has hitherto prevented the use of the method for any plant population. Symonides' (1979c) recently completed studies of sand dune plants in Poland provide complete life tables for *Androsace septentrionalis* for each of the 8 years from 1968 to 1975.

Using information on the number of individuals present in the population at the peak of germination, at the end of the seedling phase and at the end of the vegetative stage, the number of flowering individuals and the number of fruiting individuals (Fig. 5.1, p. 110), we can calculate the amount of mortality that occurred in each successive stage of the life cycle in each year. From additional information on annual seed production and by assuming that all new seedlings come

Fig. 5.4 Density-dependent mortality factors in a hypothetical population where they: (a) undercompensate for changes in population density; (b) exactly compensate for changes in population density; and (c) overcompensate for changes in population density. The diagonal (slope = 1) is drawn in as a dotted line.

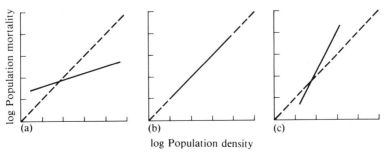

from the previous year's seeds we can also calculate the annual mortality of seeds in the soil. Variations in annual seed production which affect the number of seeds entering the seed pool each year can be calculated by comparison with the best year, giving us a mortality factor we may call 'seeds not produced'. All these data for the year 1968–69 are shown in Table 5.1.

The total mortality operating from seed production in 1968 to fruiting in 1969 is obtained from the *product* of all the mortality factors operating at each stage of the life history. To simplify the handling of these mortality data we calculate the logarithm of population size at each stage (i.e. log N_x), and may then obtain the factor by which mortality reduces the size of the population at successive stages by subtraction: $\log N_x - \log N_{x+1}$. This mortality factor is given the symbol k_x, where x is successively the stage 0 (seeds not produced), stage 1 (seeds not germinating) and so on. The k-factors have the merit that their *sum* throughout the life cycle gives us the total mortality operating in that generation. This total is called the *generation mortality* and is given the symbol K. With a complete set of k-factors for seven generations of *Androsace* we can see how important a contribution each k-factor made to the generation mortality (K) by plotting them against time (Fig. 5.5(a)). The k-factor most obviously correlated with K is k_1, the mortality of seeds in the soil. The k-factor which, on average, makes the greatest numerical contribution to the generation mortality is called the *key factor*.

The key factor, in this case seed mortality in the soil, is the strongest cause of changes in population size. It is important to realize, however, that such factors are practically never (in animal populations, so probably also in plants) density dependent and therefore they do not *regulate* the population. To find which k-factor(s) is density dependent we simply plot k_x against the log of the population size upon which it acted (i.e. log N_x). This has been done for *Androsace* in Fig. 5.5(b). The density-dependent factor turns out to be k_2, seedling mortality. (Though this is the only density-dependent factor we have found in this population, in theory there is no reason why there should be only one.) The slope of the relationship between k_2 and log N_2 is 0.2, showing that density-dependent changes in seedling mortality undercompensate

Table 5.1 The life table for *Androsace septentrionalis* and calculated k-factors for the season 1968–1969. Data from Symonides 1979c.

	Potential seed production 1968		Actual seed production 1968		Peak of seed germination 1969
Number of individuals					
N_x	$2.4*10^5$		$2.0*10^5$		1020
Log N_x	5.38		5.30		3.01
Log N_x − Log N_{x-1}		0.08		2.29	
Mortality factor k		Seeds not produced k_0		Seeds not germinating k^1	

Generation mortality K = Sum of (Log N_x − Log N_{x-1}) = Sum of k_x = 2.89

Fig. 5.5 *K*-factor analysis of the population dynamics of the sand dune annual *Androsace septentrionalis*: (a) variation in the generation mortality and its components between 1969 and 1975; (b) factors k_0 to k_5 plotted against the log of the population density upon which the mortality acted. (Calculated from data of Symonides 1979b)

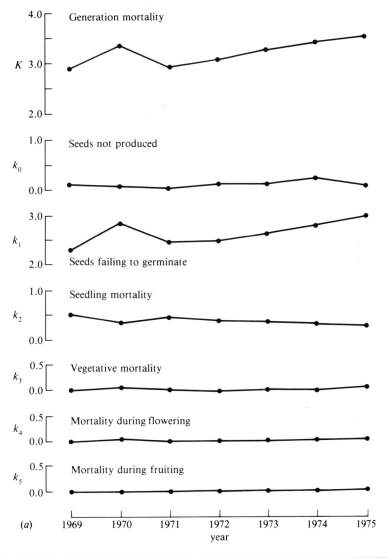

	End of seedling stage 1969	End of vegetative stage 1969	End of flowering stage 1969	End of fruiting stage 1969
	340	320	310	300
	2.53	2.51	2.49	2.48
0.47		0.02	0.02	0.01
	Seedling mortality, k_2	Vegetative mortality, k_3	Flowering mortality, k_4	Fruiting mortality, k_5

(Fig. 5.4(a), p. 113) for changes in population size. This explains why the *Androsace* population was allowed to decline in 1975 (Fig. 5.1, p. 110) when the key factor rose to a 7-year high. Nevertheless if some density-dependent reduction in seedling mortality had not occurred as population size fell, the population would have become extinct in 1975 or probably as early as 1970.

It is not easy to apply *k*-factor analysis to longer-lived organisms with overlapping generations, but this should not prevent the future useful exploitation of the method in the analysis of annual plant populations. These comprise not only some of the most easily studied plants but also many economically important crops.

Self-thinning and the −3/2 power law

The plastic growth of plants, which allows 30-year-old trees to exist as dwarfed individuals beneath the canopy of giants not much older than

Fig. 5.5b

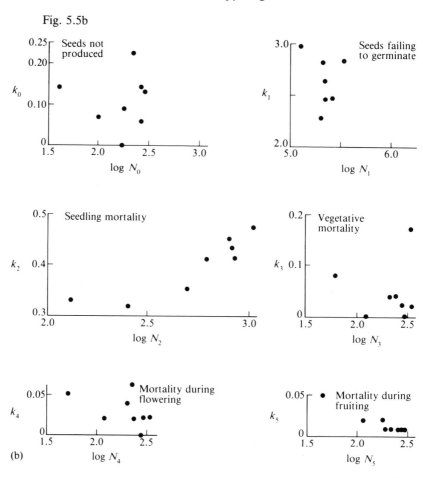

(b)

themselves, and which causes *Vulpia* to vary its seed production with population density, necessitates a special consideration of size when we try to arrive at some idea of how many plants can occupy a given space in a habitat.

If we imagine habitat space to be like a box capable of containing a given volume of children's wooden building blocks, it is obvious that such a box can accommodate a small number of large blocks or a large number of small blocks. If a number of small blocks are removed they can be replaced by a larger block of equivalent total volume: the *size* of wooden blocks and the *number* of wooden blocks are inversely related. In principle this rule also governs the number and size of plants that can exist in a population when closely packed.

While both trees and blocks obey the same rules of geometry, the analogy ends as soon as we take into account the fact that plants grow. Were plants to behave exactly like wooden blocks, a graph of log mean plant weight versus log plant density for a full habitat 'box' would have a slope of -1 (Fig. 5.6). In fact this precisely inverse relationship between the size and density of plants does apply to populations of low density which have grown to such a weight that they have reached the carrying capacity of the environment. Or, in other words, populations which have filled the habitat box to its capacity for that species.

At lower plant sizes a different relationship exists between log mean plant weight and log plant density. In populations at very low density, log mean plant weight will increase just as fast as the plants can grow and without any change in density occurring. This is obviously what will happen to the relationship between mean plant weight and density in the extreme case where only one plant is present (e.g. the sparse

Fig. 5.6 The relationship between mean plant weight and plant density expected when total plant biomass is at a maximum.

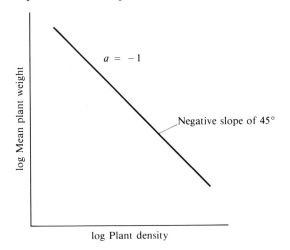

population in Fig. 5.7). Sparse populations cease to grow in *total* plant weight when they encounter the carrying capacity of the habitat, which is shown in Fig. 5.7 by a line of slope −1. Any increase in mean plant weight which occurs after this line is reached, takes place at the expense of an exactly proportionate decrease in density, i.e. at the expense of mortality.

Populations of small plants at higher densities also increase in mean plant weight as they grow, but mortality occurs before the carrying capacity is reached and before the increase in the total weight of the plant population ceases. Dense populations which have reached a size at which mortality occurs demonstrate a relationship between log mean plant weight and log density which generally has a slope of −3/2 (e.g.

Fig. 5.7 The progress of a sparse and a dense tree population through time, illustrating the main features of the −3/2 thinning process.

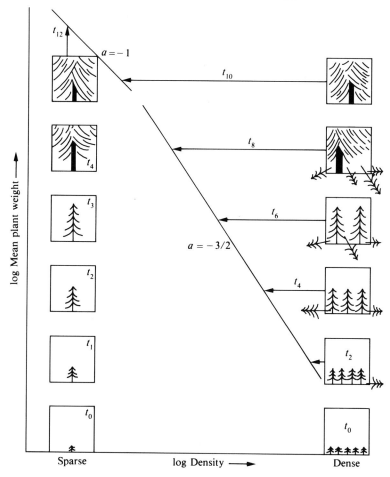

the dense population in Fig. 5.7). This means that every change of 3 units in mean plant weight corresponds to a change of only 2 units in mean plant density.

As plants in a dense population become larger with age, the density of individuals in the population decreases due to mortality. For as long as the relationship between mean plant weight and density is governed by a line with slope $-3/2$, total **plant** weight will increase because mean plant weight is increasing faster than density is falling. An example of this process, which is called *self-thinning*, is shown in Fig. 5.8. It has been observed quantitatively in about eighty species of trees and herbs whose populations show density-dependent mortality (White 1980). All demonstrate a relationship between the mean dry weight (w) of plants

Fig. 5.8 The relationship between log mean volume per tree and log density of trees in ten stands of ponderosa pine. The age of each forest stand is indicated for each point on the graph. (From White 1980)

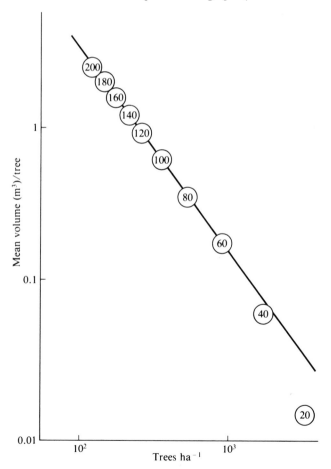

and population density (d) of the surviving individuals of the form:

$$w = Cd^{-a} \qquad [5.1]$$

where C and a are constants, a is the slope of the line in a log/log plot of density and mean plant weight and log C is the intercept of the line on the ordinate:

$$\log w = \log C - a \log d \qquad [5.2]$$

A further generalization about the self-thinning process which emerges from a comparison of different species is that C has a limited range of values between 3.5 and 4.3 (White 1980), and a frequently has a value in the region of $-3/2$ (Yoda *et al.* 1963).

The apparent constancy of the value of a led Yoda *et al.* to name [5.1] the *−3/2 power law*.

A diagrammatic summary of the self-thinning process is shown in Fig. 5.7. Its main features are:

1. Plants increase in mean size (weight) with time.
2. No density-dependent mortality occurs until populations reach the self-thinning line.
3. Mortality begins earlier in dense populations than in sparse ones.
4. Trees of the same mean weight are younger in the sparse population than in the dense one (indicated by relative *t* subscripts).
5. Both sparse and dense populations eventually reach a stage where weight increments and mortality are balanced, $a = -1$ and total plant weight no longer increases.
6. The point where $a = -1$, is reached by denser populations first.

In the space beneath the $-3/2$ thinning line is a whole set of combinations of (relatively) low plant densities and mean biomass per plant, each of which may correspond to the situation prevailing in a population at a particular moment before density-dependent mortality begins to occur. When ryegrass (*Lolium perenne*) was planted at densities of 320–10 000 plants m^{-2} by Kays and Harper (1974), they first observed genet mortality in the densest populations, followed successively by plants in less dense treatments as each increased the mean weight per genet to the point at which it intersected the $-3/2$ thinning line (Fig. 5.9). Because thinning began first in dense populations, these experienced greater overall mortality than less dense ones. This density-dependent mortality reduced the thirty-fold range of plant densities that were sown to only a six-fold range of densities at the final harvest, 6 months later. The $-3/2$ thinning line places an upper limit on the mean weight of plants in populations of a particular density, but exactly when this constraint acts depends upon how fast plants are growing and upon the initial density of the population. When the line is reached it thereafter determines just how much thinning must occur for a given increment in plant weight.

Various factors determine how fast self-thinning progresses, once it has begun. In a study of self-thinning in fleabane (*Erigeron canadensis*), an annual growing in an abandoned field at Osaka in Japan, Yoda and his associates observed a pure population decline from 122 400 plants m^{-2} to 1060 m^{-2} in 9 months. Five plots in the field were treated with quantities of fertilizer applied in the ratio 5:4:3:2:1. The process of thinning was fastest in the most fertile plots which increased mean weight per plant more rapidly than populations in less fertile plots as they raced ahead of them up the thinning line.

Although −3/2 is apparently a common value for the slope of the thinning curve (*a*) it is not universal. Kays and Harper (1974) found that when they replicated their *Lolium* experiment in 30 per cent of full daylight, *a* acquired a value of −1, indicating that the balance between growth increments and genet mortality occurred at smaller plant size than in populations of the same density grown under full daylight.

A similar result was obtained by Hiroi and Monsi (1966) for sunflower (*Helianthus annuus*) grown in reduced light intensity. Both results indicate that the value −3/2 derives from a limitation placed upon the biomass that may be sustained by the amount of light captured.

Although Yoda *et al.* (1963) originally derived the −3/2 power law

Fig. 5.9 Self-thinning in four populations of *Lolium perenne* planted at four different densities. H1–H5 are replicates harvested at five successive intervals. (From Kays and Harper 1974)

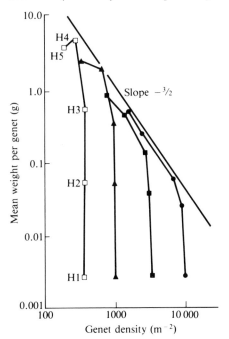

empirically, they offered a simple explanation of it derived from the ratio of plant volume to the ground area a plant occupies, and hence the resources including light available to it. To follow their explanation we will again substitute wooden cubes for real plants for the sake of simplicity. The weight w of such a plant is proportional to its volume, which is the cube of its linear dimensions l:

$$w \alpha l^3 \qquad [5.3]$$

The mean area s occupied by the plant is proportional to the square of this linear dimension:

$$s \alpha l^2 \qquad [5.4]$$

and therefore:

$$\sqrt{s} \alpha l \alpha \sqrt[3]{w} \qquad [5.5]$$

and $w \alpha s^{3/2}$ [5.6]

Self-thinning does not occur until plant density is high enough to produce 100 per cent cover. When this is reached, the mean area a plant occupies will be inversely proportional to density (d):

$$s \alpha 1/d \qquad [5.7]$$

Substituting $1/d$ for s in [5.6] we get:

$$w \alpha 1/d^{3/2} \quad \text{or} \quad w \alpha d^{-3/2} \qquad [5.8]$$

and inserting a constant:

$$w = Cd^{-3/2} \qquad [5.9]$$

This derivation of the $-3/2$ power law fails to explain certain important aspects of self-thinning. These deficiencies are discussed by White (1981).

The other major factor determining the nature of the relationship between plant size and number is seen most clearly in the between-shoot interactions in clonal plants. When such plants are considered as genets they conform to the thinning law (e.g. *Lolium* in Fig. 5.9, p. 121), but when the dynamics of the component parts are analysed, the individual shoots rarely reach sufficient density for self-thinning to operate.

Self-thinning between genets is the result of crowding, over which individual plants have no control. By contrast, the density of connected shoots produced by a single genet is dependent upon the behaviour of an individual which may control the density of ramets it produces and their disposition, by the rate of growth and geometry of a rhizome. A genet which produced ramets at such a high density that self-thinning removed some of them would presumably be at a disadvantage compared to genets which did not crowd themselves in this way. Where shoots in a clonal population do occasionally reach a self-thinning density, the evidence suggests that the $-3/2$ thinning line is not transgressed (Hutchings 1979).

The mechanics of self-thinning

While the self-thinning of dense populations proceeds in a stately fashion and with measured step up the self-thinning line, less well-understood events are taking place between individuals within such thinning populations. The reciprocal relationship between mean plant weight and the density of surviving plants has often been observed, but from the point of view of the individual plant in such a population, its progress may be marked by death or survival and a varying amount of growth. What factors determine whether a particular plant lives or dies, grows rapidly or lingers on the verge of survival?

Some idea of what happens to individual plants during the self-thinning process can be obtained from changes in the frequency distribution of individual weight with time. In the case of marigold (*Tagetes patula*) (Fig. 5.10) sown at 400 plants to a 62 cm × 62 cm box, seedlings 2 weeks old had an approximately normal distribution of weight, with some under-representation of very small plants (Fig. 5.10) (Ford 1975). The larger seedlings at this stage were distributed at random, possibly owing their extra size to faster emergence from the soil and locally favourable microtopographical conditions of the kind discussed in Chapter 2. At 4 weeks (Fig. 5.10(b)), an upper canopy began to form from the leaves of the tallest plants. These were also the heaviest individuals and were evenly distributed in the population. Though very few plants had died at this stage, a frequency distribution strongly biased towards small plants (i.e. positively skewed) had

Fig. 5.10 The changing pattern of individual plant weight in a population of marigolds, shown as the frequency of plants in twelve weight classes. Weight classes were determined by dividing the range of weights into twelve equal intervals. Size distribution at: (a) 2 weeks; (b) 4 weeks; (c) 6 weeks; (d) 8 weeks. (From Ford 1975)

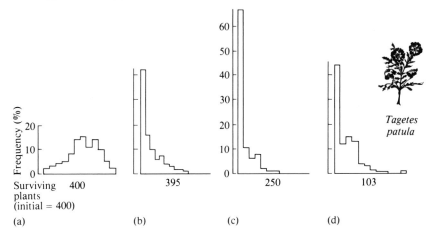

developed. Self-thinning began between 4 and 6 weeks (Fig. 5.10(c)) and selectively removed the smallest plants. Between 6 and 8 weeks (Fig. 5.10(d)) deaths occurred in a pattern which left survivors evenly distributed in space.

A similar sequence of events has been observed in the growth of a number of crop plants both in single species stands and when grown in mixtures, and also in tree populations (Ford 1975; Mohler, Marks and Sprugel 1978). These events have been described as the establishment of a 'hierarchy of resource exploitation, which results in differential growth rates among its members' (White and Harper 1970). Plants at the bottom of this hierarchy are referred to as 'suppressed', those at the top as 'dominant'. Self-thinning in pure stands of spitka spruce (*Picea sitchensis*) planted at different initial spacings shows that increased density accelerates the skewing of the size frequency distribution with time so that the hierarchy of exploitation is established more rapidly in dense stands (Fig. 5.11). In mixtures of species such as rape (*Brassica napus*) and radish (*Raphanus sativus*) (White and Harper 1970), and mustard (*Sinapsis alba*) and cress (*Lepidium sativum*) (Bazzaz and Harper 1976), resources are usually unequally divided between species so that one of them is over-represented in the suppressed class and the other over-represented among the dominants.

Among even-aged stands of trees, which generally follow the same course described for the early growth of *Tagetes*, the smallest trees may eventually be eliminated altogether, to give rise to a symmetrical

Fig. 5.11 The development of plant size distributions in populations of *Picea sitchensis* sown at three densities. Plant size is measured in twelve equal intervals of tree girth. (From Ford 1975)

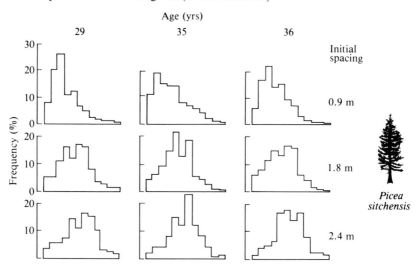

frequency distribution of sizes once again. An example of this is seen in a series of even-aged natural populations of balsam fir (*Abies balsamea*) (Fig. 5.12).

There are two ways in which the skewed weight distributions which give rise to the 'hierarchy of exploitation' in dense populations may be produced. The first explanation is contained within the terms 'suppressed' and 'dominant' used to describe the plants at either end of the hierarchy. The use of these terms implies that large plants actively interfere with smaller ones through some direct competitive mechanism. The second explanation is based upon the relationship between the relative growth-rate of a plant and its size and requires some detailed arguments.

We have already seen that populations of plants such as *Tagetes* begin growth in dense stands with a normal frequency distribution of plant weight. Relative growth rate (RGR), measured in grams per gram per day, is directly proportional to plant weight at this stage so we may assume that it too has a normal frequency distribution. When growth takes place, the additional weight produced leads to an increase in RGR. This increase in RGR accelerates plant growth, the extra growth increases RGR and so on in an exponential fashion. This process of exponential growth does not go on indefinitely of course, but it may cause plants with initially only small differences in weight to diverge very rapidly in size, with the consequence that a normal distribution of plant weight rapidly skews and becomes log normal. This is the phenomenon observed between 2 and 4 weeks in the *Tagetes* population, and it requires no suppositions to be made about interference between plants, which in any case we know is not strong enough to lead to mortality until after week 4.

Interference does occur in populations before any mortality takes place, and to what extent the argument involving RGR can explain the skewness observed in thinning populations and to what extent we must invoke the active suppression of small plants by dominant ones is a question which does not yet have a satisfactory answer. However, the observed significance of relative emergence time for the success of individual seedlings within a cohort (Ch. 2) suggests that interference is very important.

Population density and plant yield

From the point of view of a forester or an agronomist, the weight of plant material that can be obtained from a unit area planted at a given density is of more interest than the relationship between mean plant weight and density which the $-3/2$ thinning law describes. If we were buying a box of wooden blocks that were sold by weight and we were interested in value for money, we would be more interested in the

Fig. 5.12 Frequency distributions of trunk diameter for *Abies balsamea* stands divided into twelve equal intervals. Stands densities (d) are in units of trees m^{-2}. (From Mohler, Marks and Sprugel 1978)

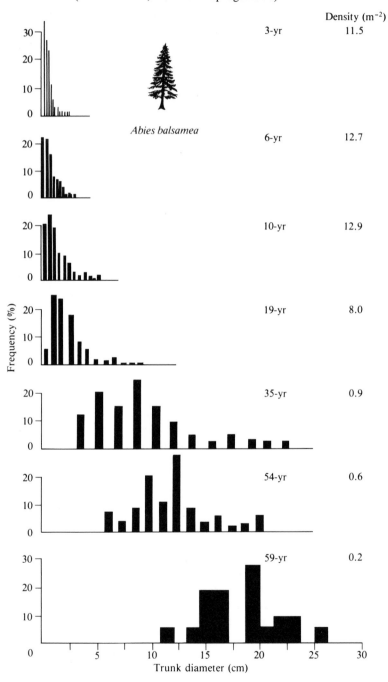

weight of blocks a box contained than in the mean size of those blocks. In our analogy the weight of blocks in a box is equivalent to the weight of plants in a unit area, or in other words the plant yield.

The relationship between yield and density in a thinning population is easily derived from the $-3/2$ thinning law. Since yield equals mean weight per plant times plant density, a yield corollary of the $-3/2$ thinning law which describes the relationship between yield and density in a thinning population can be derived as follows: The $-3/2$ law is:

$$w = Cd^{-3/2}$$

Yield = mean weight per plant $\times$ plant density:

$$y = w \times d$$

so:

$$w = y/d$$

Therefore by substitution:

$$y/d = Cd^{-3/2}$$

and cancelling both sides:

$$y = Cd^{-1/2}$$

For a population with a self-thinning line of slope $-3/2$ and intercept C this equation tells us the yield to be expected from a self-thinning population of density d. Yield/density equations for self-thinning populations with slopes shallower than $-3/2$ can be derived from the self-thinning law in the same way. For populations which have reached the upper part of the self-thinning curve where the slope achieves the value $a = 1$, we obtain a yield/density relationship:

$$y = Cd^0$$
$$= C$$

In such cases where the slope of the thinning line is -1, self-thinning populations of all densities will have the same yield, C.

Population densities which are high enough to bring about self-thinning are too high as far as a farmer is concerned because the casualties of a self-thinning crop are of no economic value. Indeed, they represent a cost in wasted seed. The same is not necessarily the case for the forester who may thin a tree stand artificially, producing a market-able crop from the thinnings before the remaining trees have reached their intended commercial size.

Two types of simple relationship between plant yield and population density have been observed for crop plants. A straightforward asymp-totic relationship is found in various crops where the yield is measured in terms of whole plant weight or some vegetative part of this, such as the roots of a beet crop or the tubers of a potato crop (Fig. 5.13(a) and (b)). These two crops illustrate two further points of interest: 1. the yield of beet roots and beet tops respond in a parallel fashion to density;

and 2. seasonal variation may affect the density at which maximum potato yield is reached. In 1956 planting densities higher than 10^4 parent tubers per acre achieved no higher yield of potatoes, but in 1958 a sowing of twice this density increased yield by about 30 per cent (Fig. 5.13(b)).

A second pattern of yield/density relationship is observed in crops grown for grain or seed. Grain or seed yield in crops such as maize (Fig. 5.13(c)), wheat, barley, soybeans, peas, ryegrass and subterranean clover grown for seed have all shown a parabola-shaped

Fig. 5.13 Yield density relationships in four crops. See text for further details. (From Willey and Heath 1969, after various authors)

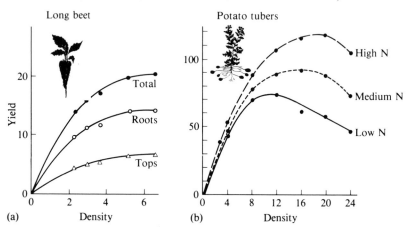

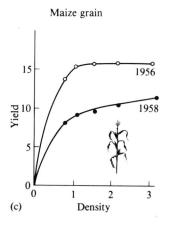

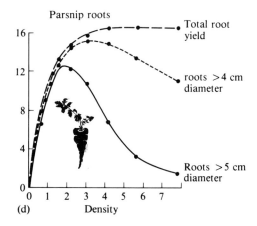

yield/density curve in experiments. In these crops, *total* plant weight is usually asymptotically related to density, but the yield from *reproductive parts* of the plant increases to a maximum and then actually decreases again as population density is raised above this. The density at which maximum seed yield is attained may vary with the nutrient status of the plants (Fig. 5.13(c)).

If the grower is interested in producing a graded crop of vegetables, only plants above a certain size may be regarded as contributing to yield. In this case even though total plant weight may be asymptotically related to density, the selected yield forms only a fraction of this and is bound to decrease in quantity as fewer and fewer plants reach the desired size. This situation, which is shown for a parsnip crop in Fig. 5.13(d), although artificial, has a relevant parallel in natural herb populations in which the probability that a plant will flower is related to its size. Many of these plants, like parsnip, are monocarpic perennials with a tap-root used for storage (p. 20).

In general, increased density may reduce seed production in various ways, either through mortality before flowering (e.g. *Androsaca septentrionalis*), by increasing the number of survivors which remain vegetative (e.g. *Digitalis purpurea*) or by altering the number of seeds produced per plant (e.g. *Vulpia fasiculata*) or by a combination of these effects. The relative importance of these different density effects appears to depend upon the species and upon the range of densities being considered. At low densities plastic changes in plant growth are probably the most important cause of reductions in most populations' seed yield.

Plastic responses to density may affect the ultimate weight and number of seeds produced by various routes. In the corn-cockle (*Agrostemma githago*), which is an annual weed of cereal crops which was once common in Britain and is still found in cereals in continental Europe, flowers are borne on variable numbers of branches, these may give rise to variable numbers of seeds capsules and capsules may contain variable numbers of seeds. In experiments in which Harper and Gajic (1961) sowed this plant in pure stands at densities of 1076, 5380 and 10 760 seeds m^{-2} they found that the yield of seeds increased asymptotically with density. As density increased, the number of seeds per plant was reduced most by a fall in the number of capsules per plant and far less by a reduction in the number of seeds per capsule. Mortality was insignificant at the densities studied and seed weight conformed to the generalization made in Chapter 4 and remained virtually unchanged over the tenfold range of plant density. In a similar experiment by Puckridge and Donald (1967), the number of seeds per plant produced by wheat grown at a twenty-fivefold range of densities was more greatly affected by a decrease with density in the number of ears per plant than by a decrease in the number of seeds per ear. The mean weight of wheat grains remained practically constant in this experiment too and no mortality occurred over the tenfold range of densities.

The manner in which reproductive output is adjusted with density depends upon the morphology of the plant. Both corn-cockle and wheat possess a structure of branches or tillers which can be multiplied during the growth and development of the plant as the resources available to it allow. When resources are scarce due to high density the first economy a plant makes is in these basic components of plant structure. The situation is different in another annual species, the cultivated sunflower (*Helianthus annuus*), which has but a single stem terminating in a single capitulum (a disc-like flower head). In some experiments by Clements, Weaver and Hanson (1929), this species was grown in a series of stands in which plants were spaced at intervals of 5, 10, 20, 40, 80, and 160 cm. Various dimensions of the plants were measured during the growth of the stands. Over the eighty-fold range of densities, mean above-ground dry weight varied by three orders of magnitude (2 g/plant to 491 g/plant), leaf area varied by four orders of magnitude (41 cm^2 plant to 27 192 cm^2/plant), but mean plant height varied only twofold, from the plants spaced at 5 cm intervals which were 101 cm high to those 160 cm apart which were 218 cm.

In the early weeks of the experiment plants at higher densities were actually taller than those at lower densities. This early difference in height in response to density occurred during the period when the capitulum was developing from the apical meristem. The additional height growth among plants at high density contrasts strongly with the effect of density on seed production and seed weight, both of which were drastically reduced in these plants. Seed number per capitulum varied by three orders of magnitude (15 seeds/capitulum to 1803 seeds/capitulum) and the weight of individual seeds by one order of magnitude (0.009 g to 0.059 g). Having only a single capitulum, sunflower adjusts its seed production in the only way it can, by reducing seed number and by sacrificing seed weight. Plants from wild populations of *Helianthus annuus*, unlike their cultivated derivatives, are branched and respond to density by reducing the number of branches and the number of capitula (Harper 1977).

The adjustments plants make in their structure in response to density, discussed here mainly in terms of reproductive organs, illustrate nicely how plants' modular construction allow the most economical size adjustments to be made. This manner of adjusting size is most evident in clonal plants which spread vegetatively.

Summary

Population *regulation* operates via *density-dependent* processes of mortality and fecundity. These processes ensure that populations remain relatively stable in numbers, despite environmental fluctuations. The various factors which determine mortality at successive stages in the life cycle may be analysed by *K*-factor analysis.

A reciprocal relationship exists between plant size and density. As plants grow, mortality occurs and plants generally obey the relationship between log mean plant weight and log density known as the $-3/2$ power law. At high mean plant weight, mortality and growth are balanced and no further increase in yield occurs.

As self-thinning progresses, the frequency distribution of individual plant weight skews and a hierarchy of exploitation is established. High initial plant density speeds up the skewing process. In mixed species stands, the species may be unequally represented among the dominant and suppressed plants in the hierachy.

A relationship between yield and density may be derived from the $-3/2$ power law. For whole plants, yield is usually asymptotically related to density, but the relationship for some plant parts shows there to be an optimum planting density to achieve the highest yield.

6
Vegetative propagation and clonal growth

Although the size of a plant is much less precisely determined by its age than is the case in many animal species, for instance in ourselves, there is more constancy in plant structure than is at first apparent when a large plant and a small one are compared. Most plants, like wheat or corn-cockle, are composed of a variable number of basic morphological units. A small grass plant is one with only a single or a few tillers to its name, a small dicot herb may have only a few branches. Larger specimens of these plants have more tillers or more branches. The tillers and branches of a larger plant may be, and often are, larger than those of a smaller plant, but this size increase is insignificant by comparison with the almost indefinite multiplication of tiller or branch *number* which may take place when a plant grows. A single tiller or branch may be described as a *module* (J. White 1979).

Collections of these modules, linked together, may form whole plants differing in size from each other according to the number of modules from which they are constructed. Where modules, like the tillers of grasses or sedges, are formed laterally and they have their own roots, they effectively form units of vegetative propagation, in other words ramets.

The regulation of clonal populations

The modular construction of plants permits most of them to be cloned by the removal of ramets which will then grow independently. This is a very common practice in horticulture, and some crops such as the banana are perpetuated entirely by this method. Vegetative propagation is also important in natural populations of wild plants. It has been estimated that more than two-thirds of the commonest perennials in the British flora show pronounced vegetative propagation (Salisbury 1942). Even greater proportions of woodland and aquatic herbs show this habit. The potential efficiency of vegetative propagation in aquatic herbs has been demonstrated by the spread of Canadian pondweed (*Elodea canadensis*) after its introduction into Britain at Market Harborough in 1845. From this introduction the species rapidly colonized most of Britain's waterways entirely by the proliferation of ramets which severed, spread and further multiplied. The species is almost exclusively female in Britain and sexual reproduction is very rare or does not occur.

Many plants have structures such as the stolons of strawberries or the rhizomes of bracken which carry ramets into the space around the mother plant. The fact that these vegetative daughters may lead an independent existence has lead to the widespread description of this process as *vegetative reproduction* (e.g. Abrahamson 1980). Since the daughters of 'vegetative reproduction' are not only identical to their mother but are often actually physically and physiologically connected to her, the use of the term *reproduction* seems inappropriate in this context and we will use the term *vegetative propagation* instead.

There are examples of every degree of separation between mother and daughter from the tillering of grasses such as *Festuca rubra* in which tightly packed clumps of ramets are formed, to the spreading habit of white clover (*Trifolium repens*) which invades the surrounding turf with stolons which widely disseminate the shoots that arise from them. The mother shoot has a short life and the combined effect of these maternal deaths, the birth of new shoots and the stoloniferous habit is that white

Fig. 6.1 The changing pattern of grass and clover (*Trifolium repens*) distribution within a pasture over a 1-year period. Stippled areas are grass-dominated, blank areas are clover-dominated. (From Lieth 1960)

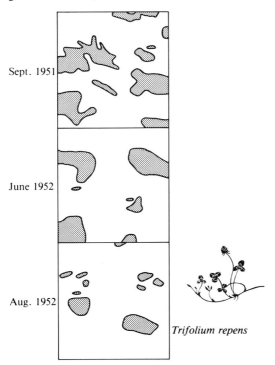

Sept. 1951

June 1952

Aug. 1952

Trifolium repens

clover wanders about in a grass sward or advances centrifugally forming
rings of clover in the turf (Fig. 6.1).

Few experiments on the density-dependent regulation of vegetative
propagation have been performed, perhaps because the physical in-
tegration of a clone makes it relatively difficult to manipulate the density
of ramets in it and to observe the outcome of these density changes. The
component shoots (ramets) of a genet rarely reach densities sufficient
for self-thinning to occur. It seems that genets behave like the prudent
farmer and do not crowd their ramets to the point at which mortality
prematurely reaps some of the harvest (Hutchings 1979). Although it
makes sense for a compactly structured genet to regulate the production
of ramets within the clone, the same is not the case for plants with an
invasive habit, since each additional vegetative shoot is a potential
colonist of new space outside the area already occupied by the genet.
Hence *Ranunculus repens* does produce stolons which invade areas
already occupied by high densities of rosettes and density-dependent
mortality is observed among them.

It is not clear from present knowledge whether density-dependent
mortality is commonly responsible for the regulation of stoloniferous
populations or whether the density-dependent *birth* of new stolons is
more often the cause of regulation. An example of the latter type of
regulation occurred in a Canadian population of the hawkweed *Hieracium floribundum* (Thomas and Dale 1974). This rosette species, like
another *Hieracium* sp. studied in Britain (Bishop, Davy and Jeffries
1978), produces plentiful seed but very few of the seedlings observed
ever become established. The plant propagates freely from stolons
which are *only* produced by rosettes which have flowered. Natural
populations of *H. floribundum* growing at a range of densities in an
abandoned pasture showed a decrease in the proportion of rosettes

Fig. 6.2 The percentage of rosettes that flowered and the number of
stolons produced per square metre in relation to the density of *Heiracium
floribundum* rosettes. (From Thomas and Dale 1974)

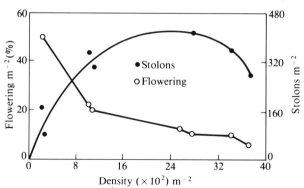

flowering with increasing density, and a concomitant decrease in the number of new stolons initiated from rosettes (Fig. 6.2). Because vegetative propagation is tied to flowering in *Hieracium* this peculiarly indirect regulation of stolon production may be unique. Nevertheless, part of the reduction in stolon production at high density was due to those rosettes which did flower producing fewer stolons each than rosettes which flowered at low density. This suggests that a more direct influence of density on the regulation of vegetative propagation also operates in this species.

The births and deaths which occur in the tiller populations of grasses reveal more about the regulation of non-stoloniferous, vegetatively propagating populations. Kays and Harper (1974) recorded tiller number as well as genet number per unit area in the density experiment described in Chapter 5. Two density-dependent processes regulating the tiller population operated during the course of this experiment: firstly, genets experienced density-dependent mortality and self-thinning (Fig. 5.9 (p. 121)) and secondly, there was density-dependent vegetative propagation of tillers by the surviving genets. At high densities the rate of genet mortality exceeded the rate of tiller formation and the overall tiller density fell. At low densities genet mortality was low and tiller formation high so that tiller densities rose. As a result of these two processes genets sown at a thirty-fold range of densities converged to a point at which all sowings had the same tiller density (Fig. 6.3). The mean weight per tiller increased through the experiment and continued

Fig. 6.3 Changes in the density of genets (continuous line) and tillers (dotted line) with time in populations of *Lolium perenne* sown at four starting densities. (From Kays and Harper 1974)

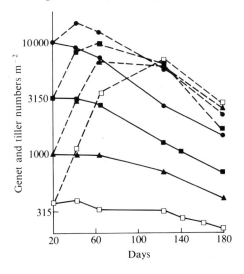

to increase once all tiller densities had converged. A log × log graph of tiller density against mean tiller weight (Fig. 6.4) shows that the increase in the mean weight of tillers was associated with a decrease in tiller density in the final phase of the experiment. This indicates that self-thinning may be detected among the parts (tillers) of genets as well as among the genets themselves.

The experiment by Kays and Harper is a very powerful demonstration of the role of vegetative propagation in the regulation of tiller density in a grass sward. An experiment of a different design by Langer, Ryle and Jewiss (1964) using two other grass species, timothy (*Phleum pratense*) and meadow fescue (*Festuca pratensis*) and some field observations of tiller demography in mat grass (*Nardus stricta*) (Perkins 1968), indicate the generality of the control of grass tiller populations by the regulation of vegetative propagation. In the experiment by Langer and his associates large concrete containers were separately sown in August with seeds of *P. pratense* to give a seedling density of approximately $4\,300\ m^{-2}$ and with seeds of *F. pratensis* to give a seed density of about half this. The swards were cut periodically and the survival of plants and the births and deaths of tillers were followed for 3 years.

In the first 6 months of the experiment a rapid increase in tiller density and an accelerating mortality of genets was observed for both species. By 12 months the rapid changes of the establishment phase had slowed down and a regular annual pattern of tiller birth and death developed. Most of the flux in tiller numbers and most genet deaths occurred in the period of most active growth from April to July (Fig. 6.5). Tiller

Fig. 6.4 The changing relationship between mean tiller weight and tiller density (both plotted on log scales) in four populations sown at different initial densities. H1 to H5 are replicates harvested at five successive intervals. (From Kays and Harper 1974)

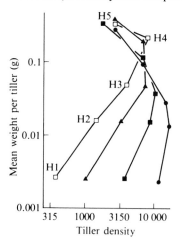

Fig. 6.5(a) The number of surviving genets per square metre in *Phleum pratense* (broken line) and *Festuca pratensis* (continuous line) populations planted in experiments by Langer, Ryle and Jewiss (1964).
Fig. 6.5(b) The number of tillers per square metre in the populations of *Phleum pratense* (broken line) and *Festuca pratensis* (continuous line). (Redrawn from Langer, Ryle and Jewiss 1964)

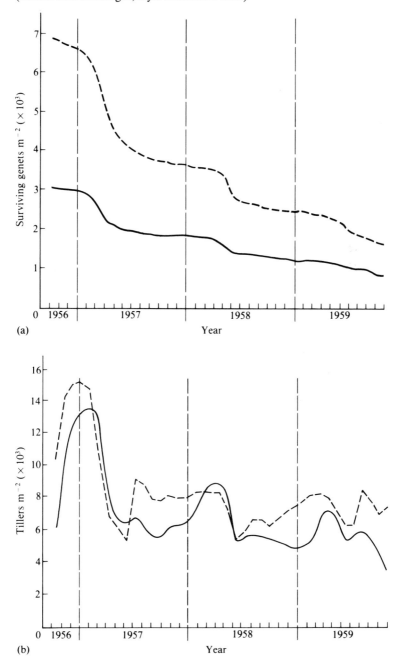

recruitment and death continued at a low level during the rest of the year too, but genet deaths were almost exclusively confined to this period. Apart from the flushes of tiller recruitment and death during the growing season, overall tiller densities remained approximately constant after the establishment phase because genet deaths were offset by the vegetative propagation of tillers from the survivors. As a result of annual genet mortality and tiller production of surviving genets, the average number of tillers per plant rose over the 3 years of the experiment. Langer and his associates summed up the changes they observed in these experimental grass swards: 'From a community composed of many plants (genets), each bearing very few tillers, the swards were gradually transformed into populations of relatively few, multi-tillering plants.' This statement also describes the outcome of tiller and genet flux in Kays and Harper's experiments.

Changes in tiller numbers within individual tussocks (clumps) of *N. stricta*, a grass of poor upland grazing areas in Britain, also demonstrates that dynamic processes of tiller birth and death regulate the size of tiller populations within a tussock. The morphology of *Nardus* tillers makes it possible to assess their age from the relative number of dead leaves and from the relative position of tillers on the rhizome. Using this method, similar to the one used by Callaghan on *Carex bigelowii* (p. 12) Perkins (1968) arrived at a population structure of five stages for *Nardus* tillers. Samples of tussocks in a vigorous state of growth were removed from the field at monthly intervals for 16 months and each sample was separated into individual plants. These were dissected into their component tillers which were then classified in the appropriate stages. These data can be used to construct a stage structure (equivalent to an age structure) for the tiller population of individual genets at each sampling date. By comparing the number of tillers present in each stage at successive dates it is possible to deduce the number of new tillers produced, the number of tillers which have moved from one stage to the next and the number which have died. This dynamic interpretation of the monthly *Nardus* tiller stage structures recorded by Perkins is shown diagrammatically in Fig. 6.6.

The diagram is drawn in circular form because the flux in tiller numbers over a 12-month period appeared to produce constancy out of change. There was heavy recruitment of new tillers in the period July to December and some deaths and movement of tillers between stages in most months, but the final number of tillers at each stage in the second December of Perkins' study was about the same as it was in the first December. This argues strongly that the tiller populations of individual *Nardus* genets were being regulated. Unlike the experimental populations studied by Kays and Harper and Langer, Ryle and Jewiss, this *Nardus* population was not artificially sown and it was certainly long established. In a situation such as this, genet mortality had probably virtually ceased, and so it is not very surprising to find population regulation operating primarily on tillers.

The fact that a plant may continue propagating vegetatively, maintaining or increasing its size by the production of ramets after genet mortality in a population has ground to a halt, suggests that such plants may have found the secret of eternal life. Indeed, there is evidence that many clonal plants are effectively immortal.

Fig. 6.6 The annual cycle of tiller recruitment and transitions between stages of tiller development and death in clumps of *Nardus stricta*. Reconstructed from monthly censuses of the tiller stage structure of *Nardus* growing in a natural population. No data were collected in March. (Drawn from data of Perkins 1968)

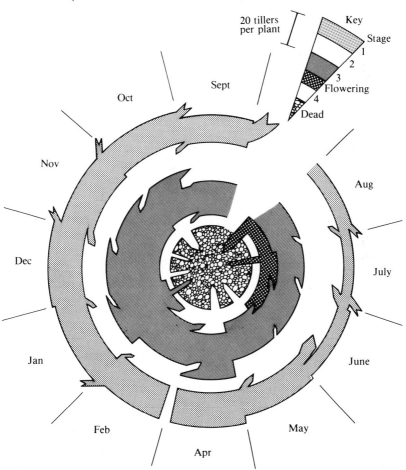

The establishment and persistence of clonal populations

The tendency towards a reduction in the number of genets in a population, accompanied by an increase in the number of ramets per genet in the remaining plants which was observed in *Lolium perenne*, *Phleum pratense* and *Festuca pratensis*, is carried to its logical conclusion in stable habitats. Clones of sheep's fescue (*F. ovina*) up to 10 m in diameter have been found in hill pastures in Scotland (Harberd 1962). In a study of another fescue *F. rubra* which grows in the same habitat, 1481 plants were gathered within a 90 m × 90 m area and identified to individual clones on distinctive morphological characters. Most of the plants in this area belonged to one of only a few large clones. In a wider sample of *F. rubra* at this site one clone was found distributed over an area more than 200 m in diameter (Harberd 1961).

Clones of this size may be the product of centuries of vegetative propagation. Their age can be determined from rates of radial spread and the diameter of clones or, in some rare cases, from historically dated events which correlate with the initiation of new clones. The latter method has been used to date the establishment of bracken (*Pteridium aquilinum*) clones in Finland. New clonal populations of this fern are established from gametophytes in open areas such as heathland. Although bracken spores are ubiquitous, they appear only to produce fertile gametophytes in burnt areas, possibly because burning 'sterilizes' the site in some way necessary for their establishment (Oinonen 1967a). Once establishment has occurred, bracken clones are resistant to further minor fires. In an extensive survey of heathlands in Finland, Oinonen (1967a, b) identified a large number of distinct bracken clones of various sizes. In areas with small clones the date of the last fire was determined from the age of nearby trees which had also established after the fire. In areas with large clones, historical records provided dates for battles and other incendiary events.

When the date of historically or dendrochronologically dated fires is plotted against the diameter of bracken clones at the same site, a remarkably close, linear relationship is seen (Fig. 6.7). At some of the same sites where Oinonen studied bracken, he found several other clonal plants which also appear to have established after fires. The diameter of clones of the clubmoss *Lycopodium complanatum*, lily of the valley (*Convallaria majalis*) and the grass *Calamagrostis epigeios* all show a close correlation with the date of major fires (Fig. 6.7). Genets of all of these plants over 300 years old have been found.

The spectacular ages reached by clonal plants emphasizes the rarity with which establishment from seed may occur in these species. Though extremely rare, the establishment of clonal plants from seed *is* important for the very reason that when it does occur it lays the foundation for a whole dynasty of vegetatively propagating genets. A genet which passes successfully through the establishment phase can look forward to a very long life, possibly with several opportunities to contribute seed and to establish progeny on the rare occasions when a new disturbed site may arise. The idea that natural selection acts heavily in the establishment phase of clonal populations, and that prolonged competition between different clones over several years allows only those best adapted to local conditions to survive, has been put forward by several writers (e.g. Harper 1978; Abrahamson 1980). This hypothesis suggests

Fig. 6.7 The diameter of clones of *Pteridium aquilinum*, *Convallaria majalis*, *Calamagrostis epigeios* and *Lycopodium complanatum* in relation to their age, as determined from the dates of fires associated with historically recorded events. (From Oinonen 1969)

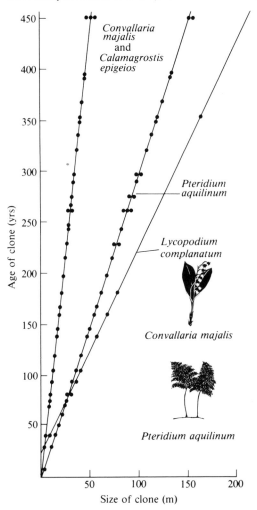

that genetic differences should be found between clones occupying different micro-environments within a habitat.

Some evidence for this was found by Harberd (1963) in a white clover population growing in sand dunes near Edinburgh. Two abundant clones, A and C, were identified and each was found to be distributed in a number of separate clumps. Clone C occurred in a number of places in a narrow zone around the rim of a dune hollow and clone A occurred at several positions within the hollow itself. In this case it appears that the extent and distribution of particular clover clones was not determined by their age but by ecological limits.

An experiment by Turkington and Harper (1979) on the same species growing in a pasture in North Wales supports this conclusion. In this particular field *Trifolium repens* is found growing in patches dominated by four different perennial grasses: *Agrostis tenuis*, *Cynosurus cristatus*, *Holcus lanatus* and *Lolium perenne*. Clover plants were removed from clones associated with each of these grass species in the field and propagated to produce a bulk population of ramets from each. Clover ramets originating from each clone were then transplanted into boxes sown with the different grass species, in every combination of clone origin and grass species. In other words ramets propagated from the clover clone growing in a clump of *A. tenuis* were planted in a box sown with this species and also in a box sown with *C. cristatus*, one sown with

Fig. 6.8 The dry weight of plants of *Trifolium repens* from a permanent grassland sward, sampled from patches dominated by four different perennial grasses and grown in all combinations of mixture with the four grass species. (From Turkington and Harper 1979)

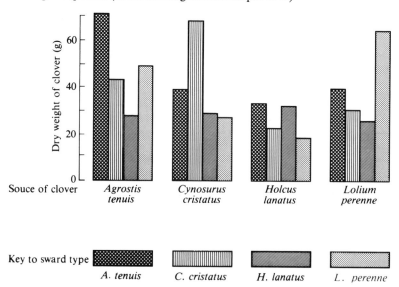

H. lanatus and one sown with *L. perenne*. Ramets from clones growing with the other grasses were treated in the same way. A year after ramets were sown into these boxes the clover was harvested, dried and weighed. The final dry weights of clover ramets from different clones planted in association with each of the four grasses is shown in Fig. 6.8. In virtually every case the clone to grow best in a box planted with a particular grass was the clone removed from a patch of that grass species in the field.

Turkington and Harper's results strongly suggest that the identity of plant neighbours is a significant factor in determining which clover clone will successfully colonize a particular patch of ground. The stoloniferous habit confers spatial mobility and vegetative longevity on *T. repens*, both of which are probably of advantage to a clone in 'finding' (on a hit-and-miss basis) its most compatible grass neighbour and persisting in that spot once it has arrived there. This is not to be taken as a suggestion that clover plants deliberately search for a home like a bird looking for a nest site, but that ultimately competition from other clones allows each clone only to persist where it is competitively superior to the others. Here we have hit upon some important features of vegetative propagation. As well as conferring potential immortality, it allows plants to explore space, to colonize favourable areas and to regulate their contacts with other clones and with other species.

The spatial organization of clonal populations

The simplest pattern of clonal growth is where a genet consisting of a single rhizome advances by increments in a straight line. A plant with this one-track approach to life, sand sedge (*Carex arenaria*), is found in sand dunes where its predominantly linear structure is put to efficient use as it invades areas of uncolonized sand (Fig. 6.9). The rhizome grows by the repeated production of a module with four nodes (leaf scars) ending in two underground buds. One of these produces a new underground module and the other may remain dormant or produce an aerial shoot with its own dormant bud. This may occasionally produce a rhizome branch. Shoots themselves go through a series of growth stages. As in the grasses discussed earlier some recruitment and mortality occurs in all months of the year but both are mostly confined to the summer months in British populations. As an area is colonized the shoot stage structure shifts from juvenile to senescent.

Rhizome buds may remain dormant long after the shoots with which they were produced have died and they may accumulate in numbers up to 400–500 m^{-2} (Noble, Bell and Harper 1979). This pool of dormant buds, analogous to the pool of dormant seeds found in other species, may form the basis for recruitment to the shoot population where new sand accumulates over an area. In a Welsh population of *C. arenaria*

Fig. 6.9 The invasion of a plot by *Carex arenaria*. (a) New shoots; (b) old shoots. The relative proportion of old and new shoots is shown next to each annual chart of the study plot. (From Symonides 1979b)

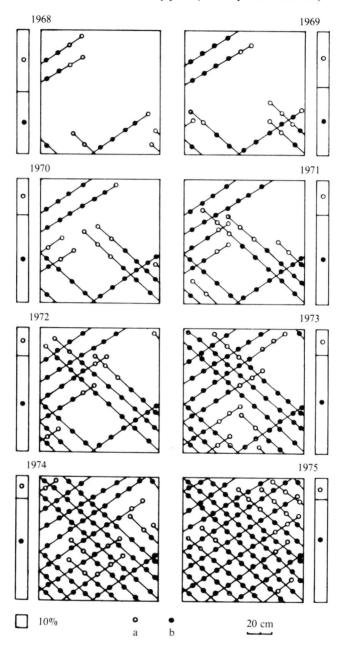

studied by Noble, Bell and Harper (1979), fertilizer was applied to some sites containing buried rhizomes with dormant buds. The flux of aerial shoots was compared in these and in untreated sites. Both shoot birth- and death-rates rose in the fertilized populations, but the net shoot population also increased substantially. The age structure of fertilized populations was rejuvenated as rapidly senescing older shoots were replaced by new ones.

The linear rhizome structure of *C. arenaria* seems particularly well suited to the marginal invasion of mobile sand. It has been suggested by Bell and Tomlinson (1980) that other specific rhizome arrangements also permit plants of other habitats to exploit the space available to them in the most efficient way. Indian cucumber (*Medeola virginiana*) is a woodland herb of deciduous forests in eastern North America whose rhizome structure allows it to 'explore' its surrounding space in the most parsimonious way. The basic module from which this plant is constructed is a straight slender stolon with a dormant bud at one (proximal) end and a tuberous swelling at the other (Fig. 6.10) (Bell 1974). The plant overwinters in this form and in the spring produces an aerial shoot and a new stolon from the tuber. By summer the proximal bud on the original stolon has broken dormancy and has produced, at an angle of 45° to the original stolon, another new but shorter stolon with a bud at this proximal end and terminating in its own tuber. The original

Fig. 6.10 The development of a clonal population of *Medeola virginiana*. (From Bell 1974)

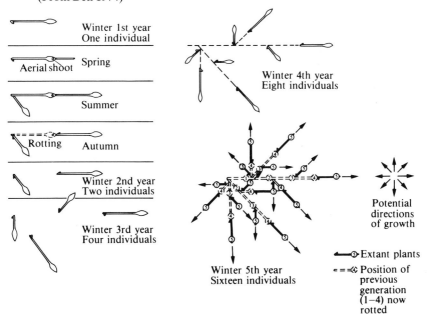

stolon, its terminal tuber and its shoot rot away in the autumn to leave the two new stolons, at 45° to each other and newly detached from each other, to overwinter. The whole process is repeated by the two new stolons in the following season and annually by their descendants, resulting in the centrifugal spread of fragments of the genet as stolons advance in all directions along radii spaced at 45° intervals (Fig. 6.10). The spaces between these advancing radii are invaded by the lateral stolons which develop from the buds at the proximal end of all stolons. The repeated production of these lateral stolons at 45° angles may, if a site is favourable, produce local plants which chase their own tails around an octagon whose circumference they complete exactly once every 8 years.

One might sum up *Medeola's* growth pattern as an answer to a basic fact of life: you can't be everywhere at once. The *Medeola* solution to this problem appears to be: at least you can have been everywhere eventually. There are examples of plants with regular branching angles in many unrelated families (Bell and Tomlinson 1980), and among them one of the commonest angles appears to be 60° (e.g. *Solidago canadensis*, Smith and Palmer 1976; *Alpinia speciosa*, Bell 1979). This particular angle produces hexagonal rhizome arrangements which are among the most efficient geometric shapes for close-packing a two-dimensional space. It seems likely that this may be an evolutionary solution to the problem many clonal plants face which is to exploit available space to the full without crowding themselves.

Summary

Vegetative propagation is common among herbaceous perennials and is also found among shrubs and trees. The habit allows the local proliferation of a clone which may persist and spread, in some cases for centuries. The ramets composing clonal populations show processes of birth, death and turnover equivalent to those found among genets in other populations.

The establishment of a new clone is a rare but genetically important event. Experimental observations with *Trifolium repens* suggest that the local distribution of clonal genotypes may be correlated with site characteristics. The spatial organization of clones may also be adapted for an optimal utilization of space.

7

Interactions in mixtures of species

Despite the tendency for clonal plants to form stands of a single species, or even of a single genet, and despite the widespread practice of farmers and foresters in Europe and North America who attempt to cultivate monocultures, most vegetation in the temperate and tropical regions contains a mixture of species. No science of plant population dynamics would therefore be complete if it could not take interactions between components in these mixtures into account. We must have answers to such questions as: when can two species occupy the same habitat without one displacing the other? How does the presence of species A affect the growth and yield of species B? How does B affect A? And how do such effects change with the density and proportions of the species? Such questions on plant competition and coexistence are not easily answered by observation alone, and our most reliable answers come from greenhouse and field experiments.

Interference and competition

When observations of two species growing in mixture show that plants of one of the species are invariably suppressed and plants of the other grow normally, it seems natural to describe this situation as the outcome of competition, perhaps competition for light, space or nutrients. Unfortunately the word *competition* is not really precise enough for a rigorous analysis of interactions between plant species because it does not distinguish between two quite distinct effects: plastic growth and mortality. As we saw in Chapter 5, plants in crowded populations may respond to density by altering their growth, by the allocation of resources to different parts of the plant or they may die. If we are interested in whether a pair of species can coexist over a period greater than one generation, it can make a great deal of difference whether plant–plant interactions result only in plastic growth changes or also in death or the failure to reproduce.

Furthermore, interactions may be asymmetrical (affecting only one species) or symmetrical (affecting both species). The symmetry of an interaction between species will obviously affect its outcome in the long term. We will call all interactions between species in mixtures which lead to a reduction in plastic growth or survival in one or both species *interference* (Harper 1961). The term *competition* will be reserved only for those situations where interference is expressed as a reduction in the numbers of plants or in the number of their surviving offspring in both (or all) species in a mixture. This restricted definition of the term competition corresponds to the use of the word in animal population ecology (Odum 1971; Williamson 1972).

The distinction we have drawn between plant interference and plant competition is an important one because it highlights a deficiency in many studies of species in mixtures. Attention is frequently focused on only one species in the mixture; for instance, on the effect of a weed on the crop with which it is growing but not on the effect of the crop on the weed. Reciprocal effects may be important, because if a genuinely competitive interaction is found it may be possible partially to suppress a weed by planting the crop at a higher density.

Additive experiments

Many experiments on the interaction of a crop and a weed employ a simple *additive* design in which a crop planted at a fixed density is sown with a weed planted at a range of densities. In a typical example of such an experiment, Buchanan *et al.* (1980) examined the effect of two weeds, sicklepod (*Cassia obtusifolia*) and redroot pigweed (*Amaranthus retroflexus*) on the yield of cotton grown in Alabama, USA. The cotton was grown in rows 15 m long and seeds of each weed were sown into the rows to give weed densities of 0, 2, 4, 8, 16 and 32 plants per row (in separate plots). Increasing weed density brought about an increasing loss of cotton yield (Fig. 7.1). At the highest weed densities weeds probably began to interfere with each other. This may explain why cotton yield diminished more when weed densities were increased from 0 to 16 weed/row than when weed density was increased from 16 to 32 per row.

There is a problem of interpretation with an additive experiment of this kind because the effects of total plant density (crop + weed) and of weed density on the weed (or the crop) are compounded. All plots with high weed density also have high total density. In order to ascertain the effect of the weed on itself, we must vary the *proportion* of weeds in the crop while maintaining total plant density constant. Experimental designs of this sort are examined later in this chapter.

Though the additive design places some limitations on the analytical information we can obtain from an interference experiment, it does have the advantage that it can simulate accurately the real situation of a crop planted at fixed density, infested with weeds. Most weedy crops are not actually two-species mixtures but multi-species mixtures containing many different weeds. Interference between species of weed as well as between weed and crop may occur, and in practice there must be the danger that removal of one weed will simply allow another weed species present to put on further growth, rather than releasing the crop from weed interference.

Haizel and Harper (1973) investigated this problem using barley (*Hordeum vulgare*), white mustard (*Sinapis alba*) and wild oats (*Avena fatua*) planted in an additive manner in 25 cm diameter pots. The experiment was rather complex in its details, but essentially it was designed to determine the effect of removing some or all the weeds in a crop planted at 12 plants per pot and infested with 24 plants of 1 weed or 12 plants each of 2 weeds. In the agricultural situations which this experiment models, the crop is barley and the weeds are mustard and wild oats. Haizel and Harper replicated their experiment so that they could treat either barley, mustard or wild oats as the crop and the remaining species as weeds, thus generalizing the nature of the whole experiment. One-half or all the weeds were removed from pots before emergence or 3 weeks after seedling emergence, and yields of total above-ground dry weight were compared with controls from which no weeds were removed.

When barley was treated as the crop and it was grown with both weeds present, wild oats was responsible for most of the loss in barley yield. Barley yield was not depressed by the low density (12 plants) of

Fig. 7.1 The yield of cotton produced from stands planted at constant density infested with a weed at a range of densities. (From data of Buchanan *et al.* 1980)

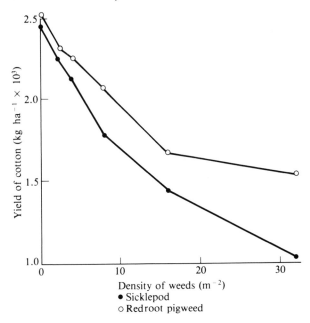

mustard remaining when wild oats were removed from the crop before emergence but barley yields were affected by a higher mustard density (24 plants). In this experiment, as in those by Buchanan *et al.* on cotton, yield loss in the crop was not linearly related to the density of weeds. When mustard was removed from a barley crop also containing wild oats, the wild oats increased its growth and almost no improvement in barley yield occurred. Where mustard or wild oats was the only weed present in a barley crop, each responded differently when half their population was removed from pots. Even when removal was late (3 weeks after emergence), the remaining wild oat plants increased their growth to compensate for the plants removed and barley yield did not improve. Barley yield was improved by removing half the mustard weed population (12 of 24 plants) but only if this was done before emergence. These experiments suggest that partial control of wild oats or control of mustard and not of wild oats (where both weeds are present) is unlikely to improve yield in barley crops.

The unusually extensive number of combinations of species and density that Haizel and Harper used make it possible to compare the effect of each species on itself and on the other species. With barley as a crop, yield was suppressed most by added barley plants, next by wild oats and least by mustard. Treating wild oats as the 'crop', suppression of yield was in the order barley > mustard > oats, and with mustard as the crop yield was suppressed in the order mustard > barley > oats. These results show that the relative effect of different species in suppressing a crop depend upon the identity of the crop. Oats have a greater effect than mustard against barley but this situation is reversed against oats or mustard.

A contrary result was obtained by Welbank (1963) in additive interference experiments in which twelve weed species were grown in populations of sugar-beet, kale and wheat. Weeds which heavily suppressed yield in kale also heavily suppressed yield in wheat, despite the considerable morphological difference between these two crops. The relative effects of different weeds in kale and sugar-beet were also similar. In experiments where two crop species of oats were grown in mixtures with four weed species of *Avena*, the relative effects of different weeds were also found to be the same on the two crops (Trenbath and Harper 1973).

Interference below ground

Weeds added to a crop or plants of one species added to a population of another introduce interference from two sources: above-ground interference between shoots and below-ground interference between roots. The below-ground component of plant interactions is often ignored, though experiments demonstrate, as reason suggests, that these are important.

An early field experiment on interference between tree roots and the roots of the ground-layer vegetation in a Scottish pinewood was made by Watt and Fraser (1933). Whereas in glasshouse or other cultivation experiments the appropriate additive experimental design involves *adding* plants in mixed culture, in the field the appropriate experiment is to *subtract* plants from existing mixtures. In Watt and Fraser's experiments they effectively subtracted the influence of tree roots from a series of adjacent plot containing wavy hair grass (*Deschampsia flexuosa*) and wood sorrel (*Oxalis acetosella*) by digging trenches around each plot. These trenches were dug to various depths so as to cut off tree roots at progressively deeper levels to a maximum depth of 46 cm. Control plots were not trenched and the tree canopy was left intact in all treatments.

The roots of *D. flexuosa* extended to about 40 cm depth at the experimental site. When compared with plants in the control, plants of this species grew progressively better in plots trenched to progressively deeper levels. *Oxalis acetosella* is shallow rooting and at this site it had roots that penetrated no deeper than 7.5 cm. In spite of this shallow rooting, this species also benefited more in deeply trenched plots than shallow ones. Watt and Fraser added distilled water to an untrenched plot and showed that this treatment did not produce the same effect on the herbs as release from interference from tree roots. This suggested that the interference effects did not involve water as a limiting factor. On the grounds of other, incomplete, evidence Watt and Fraser suspected that nitrogen might play a role in the interference between roots of Scots pine and herbs.

Glasshouse experiments on root interference between herbs have to be designed to ensure that the effects of shoot interference are not confused with the effects of roots. Somehow the roots of plants have to be added to a pot without also adding the shoots. A simple design which permits this is shown in Fig. 7.2(a).

This additive design was used by Groves and Williams (1975) to investigate interference between subterranean clover (*Trifolium subterraneum*) and skeleton weed (*Chondrilla juncea*). The latter is a major weed of cereal crops in Southeast Australia. Part of the reason for its success as a weed is due to its ability to persist through the fallow period between cereal crops when fields are sown as pasture. When subterranean clover is sown as a main component of this pasture, a 60 per cent reduction in the abundance of skeleton weed can be achieved in 4 years. One of the agents used to control the weed is a strain of the rust fungus *Puccinia chondrilliana*. The interference experiment was replicated with a set of treatments containing rust-infected plants of *Chondrilla*.

The results of the experiment at the final harvest expressed in terms of relative plant dry weight of *Chondrilla* are shown in Fig. 7.2(b).

Trifolium subterraneum suffered no significant effects of interference in any treatment and is not shown in Fig. 7.2(b). For uninfected plants, shoot interference alone had a greater effect on *Chondrilla* dry weight than root interference alone. The effect of both types of interference acting together was about what might be expected if the effects of root interference and shoot interference are multiplicative (65% root × 47% shoot = 30.6% root × shoot). Rust infection depressed dry weight to about half the control value in the no-interference treatment and by the same proportion in the shoot-interference treatment and in the root-interference treatment. Rust infection had its most severe effect in the treatment where both root and shoot interference occurred together. In this treatment the 6 per cent relative yield of *Chondrilla* was very near what is to be expected from the multiplicative effects of root and shoot interference (in rust-infected plants) (25% × 21% = 5.5%).

Replacement experiments

The most informative experiments on interspecific interference are those in which the density effects which confound the interpretation of additive experiments are ruled out, leaving only the effects of species' proportions. In these experiments the total density of plants in mixed population is held constant, but the *proportion* of the two species in the mixture is varied. In a typical experiment of this kind there might be five

Fig. 7.2 An interference experiment with subterranean clover and skeleton weed which separates the effects of root and shoot interference. (a) The four interference treatments for skeleton weed used in the experimental design. The treatments were replicated with and without a rust infection. (b) The experimental results, expressed as the dry weight of skeleton weed produced from a treatment at the final harvest as a percentage of the no-interference, rust-free treatment. (Redrawn from Groves and Williams 1975)

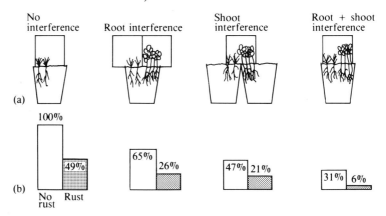

different treatments, all with the same density but with species A and B sown in the ratios 1:0, 0.75:0.25, 0.5:0.5, 0.25:0.75, and 0:1.

The yield of each species in a set of these treatments, which is conventionally called a *replacement series*, is plotted against the proportion of that species in the mixture in a replacement diagram such as the one shown in Fig. 7.3. If the sowing density at which the experiment is conducted is low, interference between plants of the same or different species may be insignificant and the yield of species A is likely to be directly proportional to its abundance in each mixture. This situation is shown as a straight line in Fig. 7.3. The corresponding line for species B is also linear, but has a slope of the opposite sign because its abundance is inversely related to the abundance of species A in the replacement series. A relative yield for each species in each mixture may be calculated from its yield in the mixture divided by its yield in the pure stand:

$$\frac{\text{Relative yield of A}}{\text{in the mixture } a:b} = \frac{\text{Yield of A in mixture}}{\text{Yield of A in pure stand}}$$

Similarly for B:

$$\frac{\text{Relative yield of B}}{\text{in the mixture } a:b} = \frac{\text{Yield of B in mixture}}{\text{Yield of B in pure stand}}$$

The sum of these relative yields for the mixture *a:b* gives us the *relative yield total* (RYT) which is a useful index of the interactions

Fig. 7.3 An idealized diagram of a replacement series in which no interference occurs between the species A and B. Yields are expressed on a relative scale. The yield of pure culture = 1.

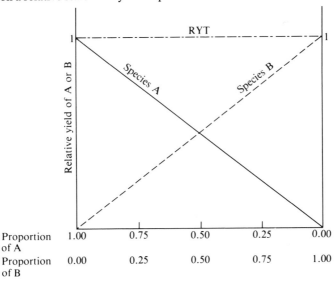

| Proportion of A | 1.00 | 0.75 | 0.50 | 0.25 | 0.00 |
| Proportion of B | 0.00 | 0.25 | 0.50 | 0.75 | 1.00 |

between A and B in a particular mixture. When the RYT for a mixture is 1 the mixture is yielding in strict proportion to the yields of pure stands. In the replacement diagram illustrated in Fig. 7.3 RYT = 1 for all mixtures in the series.

A range of replacement diagrams represents the variety of interactions that is found in this type of experiment (Fig. 7.4(a)–(d)). In diagram (a) the convex shape of the curve for species A shows that it yields proportionately more in mixtures with B than is to be expected if yield is related linearly to its proportion in these mixtures. (If the yield of A was linearly related to its proportion in mixtures the yield curve would be straight, as in Fig. 7.3.) The same applies to the yield of B in this experiment. Because the yield of both species is increased in mixtures with the other, the RYT is much greater than one. A usual biological interpretation of such a diagram is that *intra*specific interference is greater than *inter*specific interference. This could reflect differences in the use of resources such as nutrients by the two species. Some other causes of this type of interaction are discussed in Chapter 8. In general it can be said that if each species' growth is limited by a different resource, there are fewer conspecifics to compete for these in a mixture than in a pure stand and consequently both species yield well in a mixture.

Where only one species (A) benefits in mixtures and the other (B) has a linear yield (Fig. 7.4(b)), the beneficiary may be responding to reduced interference from members of its own species in mixture or it

Fig. 7.4 Replacement diagrams illustrating four types of outcome from interference in a replacement experiment. Yields are expressed on a scale of relative units with the yields of pure stands = 1. (a) Each species yields relatively higher in mixed than in pure culture. RYT in a 50:50 mixture >1; (b) A yields better in mixture. The yield of B is proportional to its density in the mixture. Maximum RYT >1; (c) A yields better and B more poorly in mixtures. Maximum RYT of a mixture may be = 1, <1 or rarely even >1, depending on the relative effects of interference on A and B; (d) Both A and B yield better in pure culture than in mixture. RYT of a 50:50 mixture < 1.

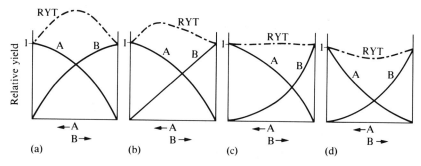

may be utilizing some resource, such as bacterially fixed nitrogen(N), derived from the other species (i.e. commensalism). Another cause of this kind of interaction is the intriguing possibility that A suffers lower predation in the presence of B because species B has chemical defences which deter herbivores (Attsat and O'Dowd 1978). It is a matter of horticultural folk wisdom, worth investigating, that carrots sown in mixture with onions suffer far less from attack by carrot root fly than when sown on their own.

Interactions which raise the yield of one species and lower that of the other in mixture (Fig. 7.4(c)) are commonly found in replacement experiments and reflect severe, asymmetric interference. The relative timing of growth by the species in a mixture is often crucial in allowing one species to capture resources and space at the expense of the other. If maize, tomato and some other crops are kept weed free in the first 30 days of growth, weeds which develop later do not significantly interfere with crop yield.

When the yield of both species in a mixture is depressed and this is reflected in their respective population sizes (Fig. 7.4(d)), the conditions of competition in the restricted sense defined earlier are met. A characteristic of this situation is that the RYT of a 50:50 mixture is less than 1.

Since we have been discussing the biological interpretation of different types of replacement diagram in a rather theoretical frame of reference up to now, it is reasonable to ask how often each of these different results is observed in actual experiments. A thorough review by Trenbath (1974) of the RYTs of 50:50 combinations in 572 replacement experiments employing mixtures of different species and different crop cultivars produces a rather surprising answer. Relative yield totals with a value of 0.9 and less (i.e. significantly less than 1) were found in under 14 per cent of mixtures, RYTs of around 1 occurred in 66 per cent of cases and about 22 per cent of experiments had an RYT > 1.1. Competition, or more correctly, mutual interference between the components of mixtures seems to be relatively rare.

Although the dearth of low RYTs suggests *mutual* interference may be rare, mixtures in which asymmetric or one-way interference occurs do not necessarily depress the RYT of the mixture (see Fig. 7.4(c)), and this type of interaction is probably considerably more common in experiments and perhaps in nature than true competition. A consideration of the conditions under which experiments are conducted will tell us why.

Are there born losers?

In our discussion of additive interference experiments we came to the tentative conclusion that certain weed species are consistently superior to others in their effect on yield when grown with a variety of crops. The identity of the weed species and the crop species that are grown in an experiment are not the only variables that should be considered before we take the serious step of conferring the title 'King of the Cabbage Patch' on a particular weed in perpetuity. Field and experimental conditions vary in more ways than the additive experiments described can take into account. Replacement series, replicated under different environmental conditions, throw more light on the context in which particular patterns of interference between plants occur and how these relationships may be changed.

Oats and barley are sometimes grown together as crops in mixture in northern Europe. Replacement experiments with these two species sown at a range of soil pH from 6.4 to 3.1 demonstrate that soil reaction has a strong effect on the interaction between these two species and that it can completely reverse the relative yielding ability of the species in mixtures. The yield of the species in the replacement diagrams for these experiments is expressed as the number of kernels produced by each species per hectare and not on the relative scale of Fig. 7.3 and 7.4. At high pH (Fig. 7.5(a)) barley yield is improved and oat yield is depressed in mixtures, though oats yield more in a pure culture than barley in a pure culture. As pH is lowered to 4.0, the yields in pure cultures remain the same but the effects of the interaction in the mixture are no longer apparent (Fig. 7.5 (b)). When lowered to pH 3.7, yields in pure culture are maintained but now oats benefit in the mixture and barley yield is depressed (Fig. 7.5 (c)). So, between pH 6.4 and 3.7 the soil reaction determines the balance of yield advantage in the interactions between oats and barley. At still lower pH, yields of both species in pure cultures suffer but barley is affected far more severely than oats (Fig. 7.5(d)). The practical absence of barley growth in the mixtures at very low pH allow oats to benefit from the extra space available: the experiment is effectively transformed into a density experiment with one species (Fig. 7.5(e)).

Replacement series replicated under different nutrient conditions or under growth conditions varied in some other way are potentially a powerful tool in the analysis of interference between plants. Because of the agronomic importance of grass/legume mixtures, a number of these have been the subjects of replacement experiments replicated in this way. Hall (1974) demonstrated that, under certain conditions, the legume *Desmodium intortum* was suppressed in mixtures with the grass *Setaria anceps*. Chemical analysis of the P, N and potassium K content of the two plants showed that the RYT values were approximately equal to 1 when yields were calculated as the dry matter produced, as the P content or as the K content of plants, but the RYT for the yield of N in a mixture was greater than 1 (Fig. 7.6 (a)). Even when its yield of dry matter was depressed , the legume was able to draw upon aerobic N that was unavailable to the grass. Relative yield totals of unity for P and K showed that both species were drawing upon the same source of P and K. When, in another treatment, K was added to the mixtures the dry matter, N, P and K RYTs all shot up and *Desmodium* was virtually released from interference (Fig. 7.6(b)). This experiment demonstrates that the availability of K controls the ability of *Desmodium* to thrive in mixture with *Setaria*, which it is able to do by drawing upon a different source of N. Phosphorus does not appear to have been a nutrient limiting the growth of either species in these experiments.

In both this experiment and the ones with oats and barley, environmental conditions appear to be able to alter the balance between species – a balance which seems to rest on a keen knife-edge. Perhaps it

Fig. 7.5 Replacement experiments with barley (○) and oats (●) performed at a range of pH values from 6.4 to 3.1. Yields are expressed in absolute units of 10^6 kernels produced per hectare. (a) pH 6.4; (b) pH 4.0; (c) pH 3.7; (d) pH 3.2; (e) pH 3.1. (Data of van Dobben, from de Wit 1960)

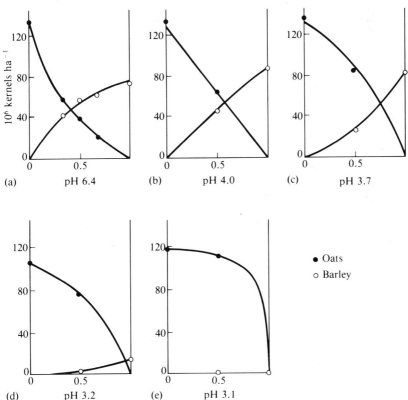

is the small difference between conditions which favour one species and conditions which favour the other which make it difficult for the experimenter to hit upon a replacement series where interference between two species is balanced and symmetrical.

Interference and coexistence

Replacement diagrams are an extremely effective means of determining the yielding performance of two species in mixtures, but this information usually applies only to plant interactions over a relatively short space of time; perhaps 1 year of growth during the establishment phase of a mixture of perennial plants, or a single generation for annual plants such as barley and oats. Just as the species with the highest yield in a pure stand will not necessarily have the highest yield of two species grown in mixture (e.g. oats in Fig. 7.5 (a)), it is not necessarily the case that the mixture with the highest yield will contain a combination of species which can persist without a change in the proportion of its component species over a long period of time. By their very nature, the long-term outcome of interactions between plants are frequently not predictable from data obtained in more simple circumstances or over only a part of the plant life cycle. Hence we need an experimental design whose specific object is to test changes in the relative abundance of two (or more) species in a mixture.

Fig. 7.6 The effect of a potassium application on the utilization of nitrogen by *Setaria* and *Desmodium* in replacement series: (a) control treatment; (b) potassium added. (From Hall 1974)

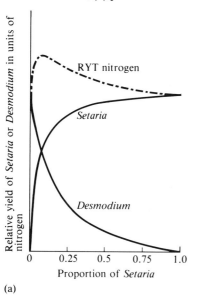

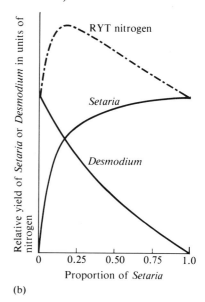

(a)   (b)

As with the replacement series, we are interested in the *proportions* of species in a mixture rather than in the effects of altering density, so an experimental design of the same general kind is appropriate. Species are sown in mixtures at fixed density but in varying proportions. At harvest(s) the ratio of plants produced (the output ratio) from mixtures with different input ratios is measured and the input and output ratios are plotted against each other on a ratio diagram such as the one in Fig. 7.7 (a). The diagonal in this diagram represents all the possible situations where a given input ratio results in the same output ratio and the mixture is stable. In the hypothetical experiment of Fig. 7.7 (a), species A increases in mixtures where it was sown in the minority (<50%) and B increases where it was sown as the minority species. The

Fig. 7.7 Input/output ratio diagrams for two hypothetical species A and B, showing four types of result. Note that input and output ratios are on a log scale. (a) A stable equilibrium which balances a mixture at a ratio of 50:50; (b) an unstable mixture in which species A increases progressively; (c) an unstable mixture in which species B increases progressively; (d) an unstable equilibrium whose outcome depends upon the initial ratio of species in a mixture. The broken diagonal line passes through all equilibrium points (i.e. where the input ratio = output ratio). Dotted lines show the trajectories followed by mixtures starting from extreme proportions of the two species.

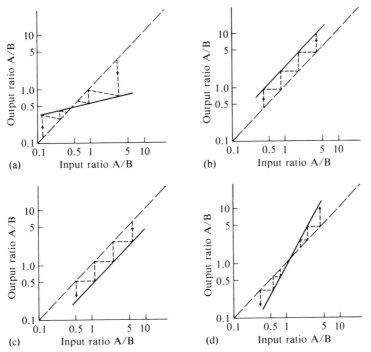

dotted lines in the diagram show how a mixture starting at a particular ratio would progress towards a stable equilibrium ratio of 50 per cent A : 50 per cent B. Figures 7.7 (b)–(d) illustrate the outcome in unstable mixtures of various kinds.

Van den Bergh and de Wit (1960) set up a replacement experiment with sweet vernal grass (*Anthoxanthum odoratum*) and timothy

Fig. 7.8 Replacement diagrams (left) and ratio diagrams (right) for interference experiments with mixtures of the two grasses *Anthoxanthum odoratum* and *Phleum pratense*. (a) The results of a field experiment; (b) the result of an experiment in a growth chamber. (Data from Van den Bergh and de Wit; diagrams from de Wit 1960)

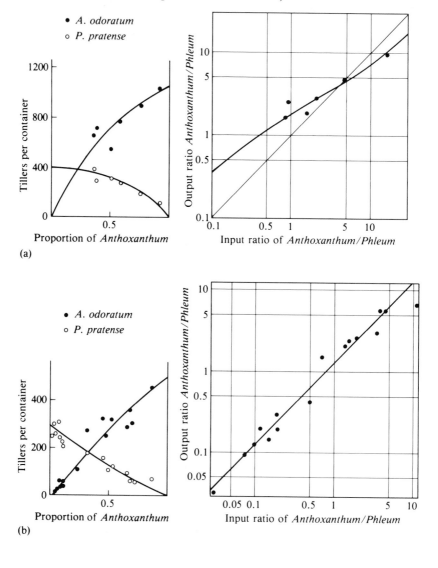

(*Phleum pratense*) in a growth chamber and also in the field. Their field results are expressed as a replacement diagram and as a ratio diagram in Fig. 7.8(a). The replacement diagram suggests that the species either mutually supply each other with nutrients (unlikely since both are grasses) or that they utilize space differently and thus reduce interference. The ratio diagram shows that a mixutre of about 4 : 1 in favour of *Anthoxanthum* is stable and that mixtures sown at greater or smaller ratios will move towards this equilibrium point. The results of the same experiment in the growth chamber produced no stable equilibrium and no sign of mutual increases in tiller number (yield) in mixtures (Fig. 5.8(b)). This difference between the results of the experiment in a growth chamber and in the field suggests an explanation for the field results.

Conditions in the field varied seasonally and within the summer season itself temperature, light and day-length changes occurred. In the growth chamber van den Bergh and de Wit simulated changes in conditions from summer to winter but they did not provide the gradually changing conditions which occur naturally over the summer season. During this period in the field the two grasses in the experiment appear to have responded differently to these changes. *Anthoxanthum* tends to develop in early summer and the particular strain of *Phleum* used develops relatively late. This difference in development times appears to have been sufficient to reduce interference between the species and to allow coexistence in the field.

The laboratory and field experiments on interference which we have reviewed suggest that some species can coexist happily side by side, while interactions between others will lead to one ousting the other. Which species is ousted will of, course, depend upon many circumstances, including the original ratio of the two species present (Fig. 7.7(d)). Within a habitat we may therefore expect to find some species occurring together more often than can be predicted from chance (positive association) and others which occur together less often than predicted (negative association).

An analysis of species' associations in several pastures in Ontario, Canada carried out by Turkington, Cavers and Aarssen (1977) confirms some of the expectations derived from experiments. Using a plotless sampling method (Yarranton 1966), Turkington and co-workers obtained a measure of the frequency of interspecific contacts for the main species in the swards which contained several different grasses, several different legumes and some other dicot herbs. Expected frequencies of interspecific contacts between all possible pairs of species were calculated from the abundance of each species, and these frequencies were compared with observed frequencies using a chi-square test (de Jong, Aarssen and Turkington 1980). A large number of statistically significant positive and negative associations between species were found from which emerged several generalizations (Fig. 7.9). No two legume species were positively associated, but most were negatively associated with other legumes and positively associated with one or more grasses. Most grasses were also negatively associated with others grasses. Dandelion (*Taraxacum officinale*) was also found to be negatively associated with two legumes (*Trifolium pratense* and *Medicago lupulina*), but positively associated with two others (*T. repens* and *M. sativa*).

These patterns of positive and negative association could be explained in two ways. The first possibility is that the environment is patchy and different species occupy different patches according to the pH, moisture content or nutrient status of the soil. Species with the same 'preference' for particular environmental conditions would then occupy the same patches and this would show up in the analysis as positive association. Species occupying different patches would also show up as negatively associated but neither type of association could be properly attributed to species interaction, since plant distribution would really only reflect the plants' responses to the physical environment and not to the presence or

Fig. 7.9 The significant negative associations between species in pastures studied by Turkington, Cavers and Aarssen (1977).

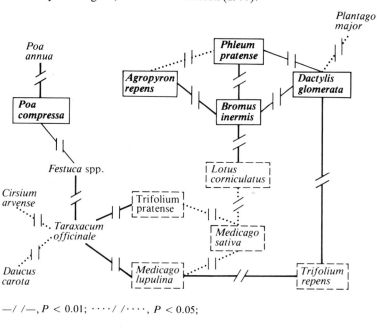

$-//-, P < 0.01; \cdots//\cdots, P < 0.05;$

[species], major grasses;

[species], major legumes.

absence of other species. The second possibility is that the patchiness of the physical environment is of secondary importance and that positive and negative association between species are the result of varying levels of interference between *particular species* pairs. If this is the case one would expect to find species with affinities for the *same* type of environmental conditions interfering most strongly with each other and consequently rarely occurring in contact.

Turkington and his co-workers attempted to differentiate between these two hypotheses as explanations for the associations between the most common grasses and legumes by sampling soil from the roots of the species. These soil samples were tested for their P, K, magnesium (Mg) and calcium (Ca) content and for their pH. No obvious relationship between these soil properties and the distribution of particular species was found. On the other hand, a statistically significant tendency was found for species that were negatively associated in the field to have the same type of soil environment. For instance *Trifolium repens* and *T. pratense* were negatively associated in the vegetation but had similar Mg, Ca and pH values in the soil attached to their roots. This result supports the idea that strong competitive interactions occur between plants with similar nutrient requirements, hindering or preventing their coexistence.

Multi-species interference and diffuse competition

There is an unfortunate discrepancy between the type of interference experiments most often performed under cultivated garden or green-house conditions and conditions in the field. Two-species interference experiments or pairwise combinations among a set of species are the simplest design for cultivation experiments. But two-species experiments are often totally impractical in the field. Field experiments unavoidably involve many species. As a rule, the simplest experiments are also the best because their results are easily interpreted. Ideally then, the simplest field experiments on interspecific interference involve removing individual species from a community and monitoring the response of the remaining species.

A number of experiments of this kind have been performed. Firstly they demonstrate that the behaviour of a plant in a mixture of two species may be quite different from its behaviour in more diverse mixtures. For instance, when plantain (*Plantago lanceolata*) was re-moved from field plots in a grassland community in North Carolina, USA, the abundance of winter annuals increased. However, this *only* happened if sheep's sorrel (*Rumex acetosella*) was absent from the experimental plot. Where *Rumex* was present and *Plantago* was re-moved, *Rumex* and not winter annuals benefited from the removal (Fowler 1981). Hence in the field situation the relationship between specific pairs of species is contingent upon the presence or absence of other species. This can lead to the apparently paradoxical result that a species may decrease in abundance when another species is removed because of the effect this removal has on a third species.

Effects of this kind plainly depend upon which species happen to occupy a particular experimental plot. The pattern of plant distribution is important. Fowler found that up to 67 per cent of the variance in the response of a species to the removal of another from her plots was due to differences in plant distribution pattern between plots.

On the whole, field experiments involving species removal demonstrate very few specific interactions between species that cannot be accounted for by the disposition of individual plants before removal was carried out. In the short term, the species to respond most strongly to the local removal of another may be whichever species happens to be nearest the gap which has been created. In the longer term, gaps may be colonized by plants regenerating from seed, and on this time-scale the size of a gap may well determine which species appears in it (see p. 38).

Removal experiments in old field communities conducted by Pinder (1975), Allen and Forman (1976) and experiments in grassland by Fowler (1981) all suggest that specific interference relationships between species are few. For instance, Pinder found that all remaining species increased their net production by about three times when clumps of the dominant grasses in the community were removed. Where specific interactions do occur they are usually not reciprocal (i.e. interference effects are asymmetric).

Interference effects on a species which derive indiscriminately from all or many of the other species in a community have been described as *diffuse competition* by MacArthur (1972). More or less weak diffuse competition between all the plant species in a community is probably a common situation, with more severe interference occurring between particular species for particular limiting factors (Ch. 8). In an experiment by Mack and Harper (1977) on interference in four-species mixtures, they were able to quantify the effects of diffuse competition on individual plants. The term 'diffuse' is perhaps misleading in the context of this particular experiment because it actually demonstrated the highly local effects of a plant's neighbours on its growth and seed production. 'Diffuse' should be understood to refer to the multi-species and multi-faceted nature of interactions, not to the spatial distribution of their effects.

Mack and Harper's experiments involved four small sand dune annuals *Cerastium atrovirens*, *Mibora minima*, *Phleum arenarium* and *Vulpia fasiculata*. The last three of these are grasses. Seeds of the four species were sown in sand in flats. Shortly after germination, seedlings were thinned to give a random distribution of plants with the four species present in equal proportions. Plants were mapped and at the end of the growing season they were harvested. The neighbourhood relationships of each plant were measured by three parameters: 1. The size of neighbours; 2. the distance separating plants from their neighbours; and 3. the distribution pattern of neighbours, determined by their location in the four quadrants of a circle around the plant. These factors were found to account for up to 69 per cent of the variance in individual plant weight. Mortality was low (5%) in this experiment, but fecundity was related to plant weight, and hence to neighbour effects, in *Vulpia*, *Phleum* and *Mibora*.

Summary

Most information about interactions between species in mixtures comes from experiments. In additive experiments, the proportion of species and total density are both varied. In replacement experiments, density is maintained constant and only proportions are varied. Experiments on root interference show that interactions below ground may be very important.

The *relative yield total* of a mixture in a replacement series and the *replacement diagram* indicate the type of interaction which is occurring. Asymmetric interference between species is most commonly observed in greenhouse replacement experiments. Various factors such as pH may alter or reverse the relative advantage of species in mixtures.

Ratio diagrams express longer-term trends in abundance in mixtures of species. The results of experiments suggest that some combinations of species are compatible and some others are not. An analysis of interspecific association in the field supports this conclusion.

Multi-species mixtures which are most often found in the field are not often simulated in experiments. Experimental removals of individual species from natural communities suggest that species have generalized effects on all their neighbours and that specific and unique interactions are rare. This kind of interference has been termed *diffuse competition*. The size, distance and spatial configuration of neighbours appear to be the most important factors determining the weight of an individual plant in a complex mixture.

8
Coexistence and niche separation

The results of interference experiments and some field observations suggest that we can expect certain species to coexist in the same habitat, while interactions between others make coexistence improbable. In this chapter we will try to uncover some of the general rules which govern the outcome of interactions between plants. What are the necessary conditions for species to coexist?

Limiting resources, niches and guilds

In all the interference experiments we have looked at, the key to a species' success in mixtures was its ability to obtain resources from the environment. These resources can be some specific nutrient which limits growth (e.g. K for *Desmodium*), or simply growing space released by the poor growth of another species under particular conditions (e.g. oats in mixture with barley at low pH). Interference between plants is intensified when the species share the same limiting resource, but interference may be reduced if this resource is divided between populations by some factor such as the temporal separation of growing periods (*Anthoxanthum* and *Phleum*). The resources a population requires to maintain or increase its size and its manner of exploiting these, when comprehensively catalogued, describe its *niche*.

A useful way of visualizing a plant's niche is to imagine it as a space with two (or more) dimensions (Fig. 8.1). To describe the niches of *Desmodium* and *Setaria* using the information from the replacement series, one dimension of this space (or axis of the diagram) is used to represent the K resource and the other represents the N resource on which the two populations depend. At certain (low) K concentrations overlap between the two species' niches is large. These are the conditions under which *Desmodium* suffers interference. When K is no longer limiting, *Desmodium* draws upon its separate source of N and the two species' niches no longer overlap. Figure 8.1 is not drawn directly from experimental results, which would have to cover a large range of proportions of N:K to provide complete information.

There is no upper limit to the number of niche dimensions that could be described for any species; we could examine the way two species respond to Ca, Mg, soil texture, temperature, etc. and plot these in a niche space of five dimensions or more. In fact the niche has been described as an *n*-dimensional hypervolume to emphasize its multi-dimensional nature (Hutchinson 1957). Although ideally one would like to be able to plot out all of the *n*-dimensional niche space, in practice this cannot be done. Instead, ecologists try to define and measure those resource dimensions which are divided with other species, since these are the dimensions crucial to the outcome of interactions between species.

Species which divide a shared limiting resource between them are described collectively as a *guild*. For example, earth mounds created by badgers in tall-grass prairie in the USA are colonized by a characteristic group of species which are not found in the surrounding, undisturbed grassland (Platt 1975). Patches of bare ground created by disturbance in chalk grassland in England also have their characteristic species (Grubb 1976). Since each of these sets is confined to a specific resource (gaps in turf) both may be called guilds. A guild is a convenient unit within which to study the partitioning of resources and niche separation between plants.

Few studies of plant guilds have been made. Among the plants found on badger mounds Platt and Weiss (1977) came to the conclusion that two of the most important niche differences in the guild were: 1. plants' different response to soil moisture content in different mounds; and 2. differential dispersal abilities which allowed plants with better dispersal to colonize mounds which are inaccessible to members of the guild with poor dispersal. Some of the general conclusions of the study were that the early arrival of some species on a fresh badger mound could pre-empt space and thus prevent other species from establishing, even when they could otherwise grow there.

In another study of six goldenrod species (*Solidago* spp.) also found in tall-grass prairie, the species occupied niches which could readily be distinguished by soil moisture variations in the habitat (Werner and

Fig. 8.1 A two-dimensional niche diagram showing the use of two mineral resources by *Desmodium* and *Setaria*.

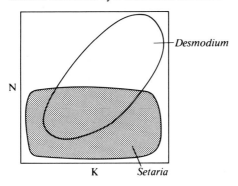

Platt 1976). The packing of species along this resource dimension in prairie and in a relatively recently colonized old field habitat where five of the species also occur are quite different (Fig. 8.2). The low degree of niche overlap in the prairie may be the result of interspecific competition eliminating plants from zones of niche overlap where interference is severe. Consistent with this is the idea that recent colonization of the field site has not allowed enough time for the outcome of competition to manifest itself in species' distributions.

Defining a plant's niche

It is clear that a species' niche is not something that can be defined entirely in terms of physical environmental factors, but that its boundaries may be set by competition from other species in regions of niche overlap and also by predators. Consideration of these factors may be incorporated into the concept of the niche if we envisage it as a large

Fig. 8.2 Percentage frequency of occurrence of individual plants in golden rod (*Solidago*) populations along a soil moisture gradient in (*a*) an old field and (*b*) in a mature prairie. (From Werner and Platt 1976)

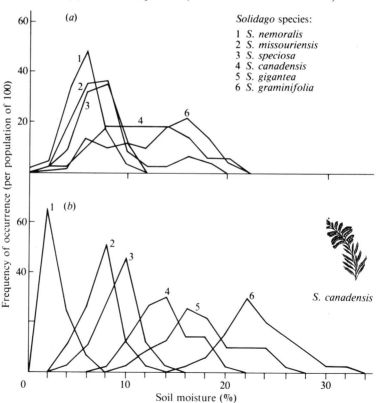

niche space incorporating all those conditions in which a population may exist when its competitors and predators are absent, containing a smaller space representing the niche of a population when its competitors and predators are present. The larger niche space is called the *fundamental niche* and the part of this which is occupied when competitors and predators are present is the *realized niche* (Hutchinson 1957).

Experiments in which competitors and predators are removed may be used to determine the extent of a plant's fundamental niche and what factors determine the limits of its realized niche. An experiment along these lines, employing two species of bedstraw (*Galium*) was performed by Tansley (1917), a full 40 years before Hutchinson's definition of the realized niche drew attention to the subject. *Galium saxatile* is a plant confined to more acidic soils in Britain whereas another species *G. sylvestre* is found in calcareous habitats. Tansley grew each species in soil from its native habitat and in soil from the other habitat in monocultures and in mixtures of the two species together. Both species grew on both types of soil when sown alone but when sown in the presence of the native species the alien species was suppressed by interference from the native and only the native species grew successfully.

It is rarely as easy to discover the factors which confine a species within its realized niche as Tansley's experiment suggests, because plants growing in the same habitat do not generally show responses to environmental variation which are as clearly defined and well differentiated as the Galiums' response to soil pH. For instance, among the plants of rivers and streams in Britain and North America, species are differently distributed according to a large number of factors of which only a few are: the speed of water flow, substrate type, width and depth of channel, distance from the mouth of the river, turbidity of the water and pH (Haslam 1978). All these factors and more, taken in conjunction with the effects of predators and interspecific interference, should be enough to deter any ecologist from trying to unravel the niche relationships of these species. Most other habitats are scarcely any more simple.

Fortunately a field method exists for cutting through some of the problems in mapping plant niches. It involves the simple technique of taking plants from sites in which they normally grow and transplanting them into parts of the habitat in which they do not normally occur. The site of the transplants and the controls should differ in some measurable or definable way likely to represent an important limiting factor in natural populations. The fate of these transplants or *phytometers* then tells us how the habitat 'looks' from the point of view of the plant itself. There is no surer way of telling whether a plant population can exist under certain conditions and whether particular factors are relevant in defining its niche than placing it in those conditions and finding out.

Phytometer populations often show qualitative changes or only gradual mortality in their new locations and several years may elapse before whole transplant populations actually become extinct. Measurements of the growth of phytometers may therefore be necessary to predict their long-term survival. Clymo and Reddaway (1972) studied a guild of four species of *Sphagnum* moss which occupy different niche positions in relation to microtopography in mires in the northern Pennines of England. They measured the growth in 1 year of phytometers of each species placed in its usual microhabitat and in the habitats of the other species (Fig. 8.3). Assuming that dry matter increase of phytometers is a realistic indicator of population growth, the results suggest that species normally occupy niche positions in which they are competitively superior to other species and that interspecific interference determines the boundary of each species' realized niche. Note that a species such as *S. rubellum* may be forced to occupy a sub-optimal part of the habitat by this interference. Species are not necessarily found in those niche positions where they grow at their *individual best*, but where they grow *relatively better* than (or as well as) other species in the guild. Another example of this is found in the bluebell (*Endymion nonscriptus*) which is commonly found in woodlands. When phytometers of this species are planted outside a wood, free from the interference of other species, they grow better than plants grown in the wood. In Britain the species is usually only found in woodlands, to which it appears to be confined by the interference of plants growing on the woodland margin (Blackman and Rutter 1950).

Oak–hickory forest in North America contains several different oak species, a number of which have local distributions which do not overlap. An effective phytometer experiment was carried out by Bourdeau (1954) to determine the factors responsible for niche differences between oak species growing at two contrasting sites in the Piedmont of North Carolina, USA. *Quercus rubra* and *Q. coccinea* grew in forest with a closed canopy on a rich, damp, Georgeville-type soil whose surface was covered in deep leaf litter. *Quercus stellata* and *Q. marilandica* grew in forest with an open canopy on poor, dry Orange-type soil with only a shallow covering of leaf litter. Trees of one soil type did not occur in forests on the other soil type.

Seedlings of all four species were planted in field plots in forest on Orange soil and in forest on Georgeville soil. Phytometers of all four species survived well on the rich soil, but the survival of the native species (*Q. stellata* and *Q. marilandica*) was superior to that of the two non-native ones on poor soil. The drought tolerance of the four species, assessed in a laboratory experiment, showed seedlings of the rich-site species to be more susceptible to drought than those from the poor site. This was therefore probably the factor responsible for the poor performance of *Q. rubra* and *Q. coccinea* phytometers at the poor site. As for

Q. stellata and *Q. marilandica*, phytometers of these poor-site species
survived well in rich forest and some other factor must account for their
exclusion from such sites. Bourdeau measured the growth of seedlings
of each species in pots of Georgeville soil placed beneath the shade
of a closed tree-canopy. This experiment showed that the poor-site
species grew little under shade and that the rich-site species grew well.
Laboratory measurements of photosynthesis in *Q. rubra* and *Q. marilandica* at a range of light intensities showed that the optimum light

Fig. 8.3 The growth of four *Sphagnum* species transplanted into native
(hatched) and non-native parts of the habitat in which they occur. (From
Clymo and Reddaway 1973)

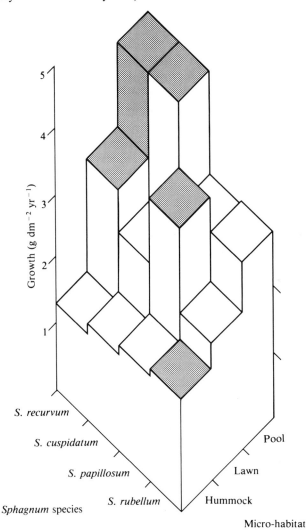

intensity for growth of the rich-site species was half that of the poor-site species. Complimentary environmental constraints, probably reinforced by competition, therefore exclude the oaks of rich sites from poor sites and those of poor sites from rich ones.

Two generalizations emerge from the results of phytometer experiments. Firstly, plants transferred to an alien part of the habitat or to an entirely new habitat may survive for quite a long time before they die and the population 'retreats', under the impact of various factors, back within the boundaries of the realized niche. Darwin pointed out the importance of interference from other plants in this process when he observed, in *'The Origin of Species'* (1859), that many alien plants (exotics) may be grown quite successfully in gardens where they are freed from interference but that they cannot survive outside gardens where they must contend with native species. The second generalization that may be drawn from phytometer experiments, particularly that of Bourdeau using oak seedlings, is that populations occurring outside their realized niche are most vulnerable to extinction in the regeneration phase of the life cycle. This is consistent with asymmetrical interference between mature plants and seedlings and the low survivorship of seedlings often observed in populations, even when growing in the native habitat (Ch. 3). Gardeners not only tend their exotic specimens by freeing them from weed interference but they often must also repeatedly sow plants in order to maintain garden populations which cannot regenerate naturally.

The vulnerability of plants in the regeneration phase and the ability of individual plants, once established, to survive despite quite severe odds against them suggests that coexistence between some species may be explicable in terms of their relative success at regenerating. An inferior competitor may obtain a counterbalancing advantage against a superior one if it is able to produce more seeds and successful progeny than the competitor or if it is able to regenerate in gaps created at a time of year or in particular years when seeds of the other species are not available (Grubb 1977; Fagerström and Agren 1979).

No studies have so far been done in which both interference between adult plants and regeneration behaviour have been compared for a pair of species. An appropriate experiment should use a replacement series to measure interference between the species. Coexisting plants, unevenly matched in competition but with relative levels of seed production which redress the balance of advantage between them, are probably to be found among the many species which colonize gaps in grasslands and in forests. It is certain that interference between trees and the herbs which spring up under gaps opened in the tree canopy by tree falls is heavily (but not exclusively) against the herbs. The latter produce large quantities of seed which allow them to colonize gaps quickly and to lie dormant in the soil between times. This reciprocal relationship of interference and relative seed production must also occur between some tree species.

Some of the different patterns of seed dormancy we discussed in Chapter 2 result in differences in the seasonal timing of germination between species in the same habitat. European arable weeds exemplify this and produce quite different floras in spring and autumn sown crops (Hanf 1973). An interesting pair of species is *Avena fatua*, which germinates in crops sown in the spring, and *A. ludoviciana* which germinates in those sown in the autumn (Grubb 1977). Hairy willow herb (*Epilobium hirsutum*) and purple loosestrife (*Lythrum salicaria*) are a pair of perennial herbs, both of which colonize gaps caused by erosion of river banks in eastern England. Seeds of both species are shed in autumn and some germination occurs in both at this time of year where seeds land on some fresh mud. Growth of *Lythrum* is slow in the shortening days and cold temperatures of autumn. Where *Epilobium* is also present in a gap formed at this time of year it quickly excludes *Lythrum* by faster growth. The relative advantage of the two species is reversed in the late spring. Gaps formed during the winter and available for colonization in the spring are filled by *Lythrum* which germinates more rapidly and grows faster than *Epilobium* in long days and warmer temperatures (Whitehead 1971). Thus the two species coexist.

Competition and coexistence in stable environments

The long-term outcome of interference between plants and the probability that they will coexist depends upon the stability of the environment. Where two species in a *stable* environment share the same niche, or in other words an identical factor(s) limits population growth in each, the two species will compete severely. This competition will virtually always lead to the exclusion of one species. This conclusion, known as the *competitive exclusion principle* or Gause's principle (Gause 1934), derives from some simple facts about population growth. Where two species compete for a resource which limits population growth in both of them, a population which has even the smallest advantage over the other in utilizing this resource will be able to increase its population growth accordingly. Even if this increase is fractional, the geometric capacity for increase inherent in all populations will magnify this advantage as successive generations of offspring are produced. This will eventually eliminate the competing population. Although circumstances may alter which of the two species wins in this situation, coexistence of two species with the same niche in a stable habitat is as improbable as the chance that their respective abilities to utilize the limiting resource are exactly identical. This chance is infinitesimal.

The situation is different in an *unstable* environment or in a habitat which is open to recolonization by propagules (e.g. seeds) from outside.

The competitive exclusion principle may not apply in these non-equilibrium conditions because its final outcome is continually interrupted by disturbance and re-establishment which sets the balance in any competitive struggle back to the beginning.

A simple mathematical model of competition based upon the logistic growth equation can be used to derive the conditions in a stable environment which will lead to the coexistence of competing species. Consider the logistic growth of a single-species population whose numbers we will measure as X. As we saw in Chapter 1 the growth of X is given by the equation:

$$\frac{dX}{dt} = r_x X \frac{(K_x - X)}{K_x} \quad [8.1]$$

where X is the population size, r_x is the intrinsic rate of natural increase of the population and K_x is the carrying capacity of the environment for species x measured in the number of individuals of x it can support. A similar equation can be written for another species, y:

$$\frac{dY}{dt} = r_y Y \frac{(K_y - Y)}{K_y} \quad [8.2]$$

Now imagine that species x and y are competitors (in the strict sense) and that the presence of some y individuals in the community reduces the number of x individuals that limiting resources can support, and that reduction is in proportion to the number of y individuals present, by a factor aY. The growth of population x is now given by:

$$\frac{dX}{dt} = r_x X \frac{(K_x - X - aY)}{K_x} \quad [8.3]$$

The term aY is a measure of the interference that population y exercises on population x by reducing the carrying capacity from K_x to $K_x - aY$. The coefficient a is called the *competition coefficient*. Population growth without competition ceases when $K_x - X = 0$, and with competition when $K_x - X - aY = 0$. Since interference between populations x and y is mutual, the equivalent equation for population y when it is in competition with population x is:

$$\frac{dY}{dt} = r_y Y \frac{(K_y - Y - bX)}{K_y} \quad [8.4]$$

where b is the competition coefficient of x on y. Population growth of Y under competition ceases when $K_y - Y - bX = 0$.

The object of this analysis is to discover when competition from one species brings population growth to a halt in the other, heralding its exclusion. In mixtures of species the population to cease growth first will obviously be the one which is excluded. The way in which the population size of x changes in relation to the population size of y can be plotted on a *joint abundance diagram* on which may be drawn an isocline which passes through all those points where the population growth of x is zero (Fig. 8.4(a)). This line corresponds to the situation in the logistic model when $K_x - X - aY = 0$. Beneath this line population X increases, above it X decreases. The intercept of the line with the y-axis is the point where $X = 0$ so the value of Y at this intercept is K_x/a (i.e. the carrying capacity of x divided by the competition coefficient of y). At the intercept of this line with the x-axis, $Y = 0$ so at this point $X = K_x$ (i.e. the carrying capacity of x in the absence of y).

An isocline for species y has been added to the diagram in Fig. 8.4(b) The intercepts for the y isocline ($K_y = Y - bX = 0$) are K_y and K_y/b. The isocline for population y is beneath that for population x in this diagram and consequently population x will continue to increase after population y has entered the region of joint abundance where it decreases. This is a competitive interaction in which y is always excluded.

Now that we have defined the intercepts of the two isoclines, we can also define the conditions which produce various outcomes. Extinction of y occurs when $K_x/a > K_y$ and $K_x > K_y/b$ (Fig. 8.4(b)). Extinction of x occurs when $K_y > K_x/a$ and $K_y/b > K_x$ (Fig. 8.4(c)). An unstable equilibrium eventually resulting in the extinction of either x or y occurs when $K_y > K_x/a$ and $K_x > K_y/b$ (Fig. 8.4(d)). Stable coexistence of two species is only possible when $K_x/a > K_y$ and $K_y/b > K_x$ (Fig. 8.4(e)). In other words, a stable mixture of two competing species will only form when each species inhibits the growth of its own population more than that of its competitor in mixtures.

We have already come across a way in which the conditions for stable coexistence may be met in the discussion of replacement series in Chapter 7. Where two populations are limited by different resources, individuals suffer greater interference from conspecifics than from plants of the other species. In such a situation the total interference suffered by a population is relatively small when a species is the minority component in a mixture, but interference increases as the frequency of the species in the mixture increases. An example of this phenomenon, known as *frequency-dependent* interference, has been observed in a set of experiments conducted with five species of poppy by Harper and McNaughton (1962).

Four of the species were annual weeds which often occur together in arable crops in Britain: *Papaver rhoeas*, *P. dubium*, *P. lecoqui* and *P. argemone*. A mediterranean weed species *P. apulum* was also used. The full experimental design included sowings of each species at a range of densities and in two-species mixtures at two different proportions (only a simplified account of the full treatments is given here). The sowings at different densities showed that plants generally suffered density-dependent mortality when sowing density was raised from 50 seeds per 35 cm $\times$ 35 cm plot (400 m^{-2}) to 225 seeds per plot

Fig. 8.4 Joint abundance diagrams illustrating the outcome of interspecific competition based on the logistic model of population growth. (a) An isocline for one species x; (b) isoclines for two species, illustrating the competitive exclusion of y by x; (c) exclusion of x by y; (d) unstable equilibrium; (e) stable equilibrium.

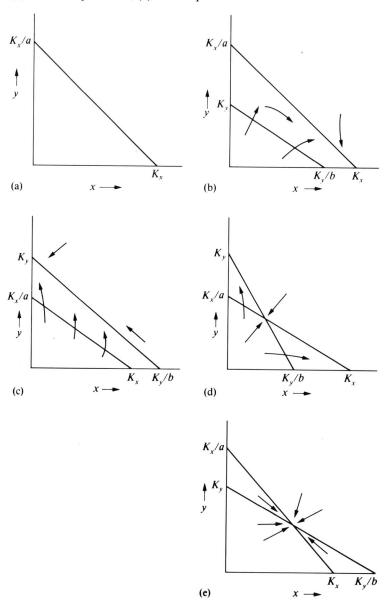

(1800 m^{-2}). The effect of a species' frequency on its survival in mixtures at the higher density was determined by sowing each pairwise combination of the five species in two mixtures. In one mixture at this density a species was sown as the majority (200:25 seeds) and in the other it was sown as the minority component (25:200 seeds). The probability of a seed producing a mature plant when sown at low frequency was then compared with the probability of a seed of the same species producing a mature plant when sown in the majority for all pairs of species (Fig. 8.5). In sixteen out of the twenty pairwise comparisons, species survived density-dependent mortality better at low frequency than at high frequency, a result consistent with these species' coexistence in the field. Three of the four mixtures where survival was not frequency-dependent involved *P. apulum*, which does not occur with other species in Britain, and the fourth mixture contained *P. dubium* and *P. lecoqui* which, of the four British poppies used in the experiment, are the ones least often found together in the wild.

Although a frequency-dependent process that could explain the ability of *Papaver* species to coexist in the wild was found in experimental populations, the mechanism was not uncovered. One possibility, already discussed, is that different *Papaver* species were limited by different nutrients. An alternative is that each species is susceptible to a specific pathogen which attacks it when the plants occur at high frequency but not at low frequency, perhaps because transmission between susceptible plants is too difficult in such conditions (Chilvers and Brittain 1972). The problem of coexistence between species is but a variation on the problem of the coexistence of genetically different strains within a population of a single species. Flax (*Linum usitatissimum*) occurs in genetically heterogeneous populations which consist of a number of strains, each susceptible to a particular strain of flax rust but resistant to those which attack other flax strains (Flor 1956, 1971). Strains of wheat which differ in their resistance to strains of the wheat bulb fly are also known. These polymorphisms and other examples in cereals may be maintained by frequency-dependent mortality caused by pests (Scott *et al.* 1980).

It is quite common for insect herbivores to attack plants in a frequency-dependent manner. Single-species stands of plants often carry a greater variety of monophagous herbivores as well as greater total numbers of them than plants of the same species isolated in a mixture of other species (Root 1973). In only a few cases have the consequences of this for the survival of plants themselves been studied.

In one such study cucumber plants (*Cucumis sativus*) were planted in plots where they were mixed with maize and broccoli and in other plots as monocultures (Bach 1980). Cucumber is the preferred food of the striped cucumber beetle (*Acalymma vittata*) whose larvae and adults both feed on the plant. The numbers of beetles on cucumbers in each

Fig. 8.5 The relative chance that a seed of a test species of *Papaver* will produce a mature plant when sown in a mixture with another species at a total density of 225 seeds per 35 cm × 35 cm plot. All pairwise of combinations of five species are shown. Open columns show the success of the test species when sown in the majority (200 seeds test: 25 seeds companion). Shaded columns show the success of test species when sown in the minority (25 seeds test: 200 seeds companion). (Drawn from data of Harper and McNaughton 1962)

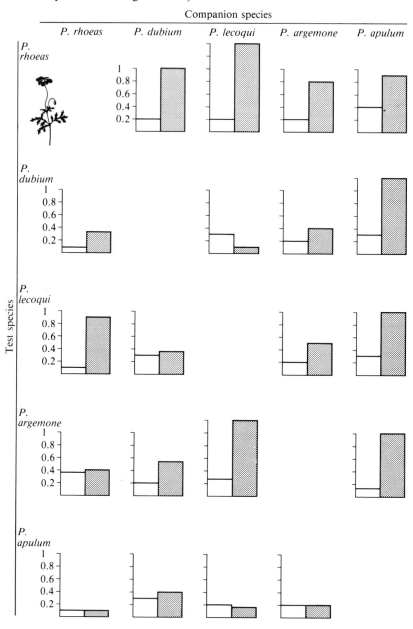

plot was counted through the growing season. Beetles stayed longer on plants and their population growth-rate was higher in monocultures than in mixed ones. The cucumber mortality rate was also significantly higher in monocultures, but surviving plants in these plots had greater fruit yields per plant than survivors in mixed culture, suggesting that a reduction in between-plant interference occurred. A bacterial wilt disease transmitted by the beetle, rather than the direct consumption of the herbivores themselves, was the most important mortality factor in the cucumber population. Quite apart from the agronomic importance of frequency-dependent insect/plant interactions, such phenomena may also be important in permitting species to coexist in more natural ecosystems.

The competitive exclusion principle faces us with a paradox in ecosystems where large numbers of plant species coexist, all apparently sharing the same environmental resources. This problem is most acute in habitats of high diversity such as occurs in some tropical rainforests where over 100 different species of tree may be found in 1 ha, with distances of up to several hundred metres separating one individual from any other of the same species. If the diversity of plants in tropical forests is high, the diversity of herbivorous insects in these forests is even higher. This may be the key to the maintenance of high diversity among the plants these insects consume.

If we consider a tree which is dispersing seeds or fruit, most of these will land near the parent (Fig. 8.6(a)). The parent tree also harbours large populations of specific insect herbivores whose own powers of dispersal are often limited, though possibly not as limited as those of the seeds. Not only a tree's seeds but also its specific predators will

Fig. 8.6 A model showing the probability of a seed (or seedling) escaping predation as a function of distance from the parent tree. (*a*) The distribution of seeds; (*b*) the probability of a seed escaping predation as a function of distance from the parent; (*c*) the product of curves (*a*) and (*b*) is the distribution of surviving seeds. (After Janzen 1970)

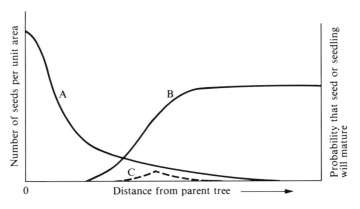

therefore be concentrated in an area around its trunk and so the latter are likely to consume any seeds or seedlings which appear within their range. A few seeds, often carried away by animal dispersal agents, will reach extraordinary distances from the parent and will find themselves beyond the reach of specific predators (Fig. 8.6(b)). These seeds are the only ones from a crop likely to survive and to be recruited into the adult tree population (Fig. 8.6(c)). Janzen (1970, 1980) has suggested that specific predators operating in this way may prevent some trees in tropical forests forming single-species stands and that this may be the important factor facilitating the coexistence of large numbers of tree species. Experiments in which the survival of seeds and seedlings placed at varying distances from a parent tree have been monitored, support the hypothesis (Janzen 1971, 1972) but some other tests do not (Hubell 1980).

A similar hypothesis to explain the coexistence of plant species and the maintenance of species diversity might be applicable in other habitats where monophagous herbivores exert a significant impact upon their prey species. This possibility has rarely been tested in temperate ecosystems. A variation on this hypothesis in which polyphagous herbivores consume the most abundant plant species and then switch their attention to other species as the abundant species becomes rare is perhaps more applicable to habitats such as grasslands grazed by vertebrates.

Competition and coexistence in unstable environments

The ecologist looking for niche differences to explain coexistence must beware of a circularity in niche theory. The competitive exclusion principle states that no two species, sharing the same niche, may also share the same habitat (i.e. coexist). There is a danger that with this principle already firmly planted in his or her mind, the ecologist assumes that the principle *must* apply in a particular case and species which coexist must *therefore* have different niches. By scrutinizing coexisting species more and more closely, examining their relative rooting depth, flowering time, pollinating agents and every character of any conceivable relevance, enough differences can often be detected to 'explain' their coexistence. Measure enough characteristics of any two objects or any two species with an accurate enough scale and you are sure to find differences between them. There are statistical methods for determining the likely competition that particular levels of niche overlap will produce (Horn 1966; MacArthur and Levins 1967; May 1975), but these really only offer circumstantial evidence that competitive exclusion will or will not, should or should not take place *if* the correct characteristics have been measured. We must use other methods to judge whether niche overlap in particular resource dimensions is significant in the dynamics of plant populations. In fact the only reliable tests are experimental ones such as those employing phytometers.

Although experimental tests are necessary to determine whether competitive exclusion is important in determining distributional patterns of particular species, some general rules which govern the applicability of the competitive exclusion principle can be derived from mathematical models. This approach was used by Huston (1979) to determine the effect of environment instability on the outcome of interspecific competition.

First, Huston simulated competition between two species using a computer model based upon the Lotka–Volterra equations. As expected, competition for resources led to the extinction of one species (Fig. 8.7(a)). Next, a density-independent population reduction (e.g. removal of 50% of each population) was introduced into the model at fixed intervals (Fig. 8.7(b)). This led to coexistence of the species for quite a while, but eventual extinction of one of them. Finally a model with six competing species was run in which the frequency of population reductions and the growth-(recovery) rate of populations could be varied. This showed that six species, sharing the same limiting resource, could coexist for a long period of time provided that: 1. all population growth rates were low; and 2. population reductions were not too frequent.

This result is extremely interesting in the light of what we know about disturbance in natural plant communities. Some of the most diverse of these, such as tropical forests, are subject to periodic disturbance of the

Fig. 8.7 (*a*) The outcome of competition between two species in a stable environment; (*b*) the outcome of competition between two species when there is a periodic, density-independent population reduction. (After Huston 1979)

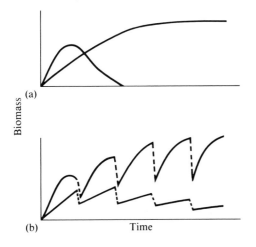

(a)

Biomass

(b)   Time

kind which Huston simulated in his model (Ch. 3 p. 70). There is evidence that periodic disturbance could also be important in maintaining high diversity in other types of habitat. Pastures freshly sown with grass rapidly acquire weeds through the colonization of molehills, and earthworm activity creates a similar opportunity for the entry of weeds into lawns. Lawns can be maintained weed free for very long periods of time by acidifying the soil so that earthworms are excluded.

We can now also see more clearly why measurements of niche overlap for *Solidago* populations in disturbed environments tend to be greater than overlap measured in stable ones. It seems that competitive exclusion between *Solidago* species in tall-grass prairie eliminates individuals from sites at which they are competitively inferior to another *Solidago* species, thus reducing the proportion of sites at which different species are to be found growing together. This process of competitive exclusion is continually interrupted in disturbed habitats and consequently niche overlap is greater in these.

Gaps in vegetation, such as are created by moles, earthworms or tree falls can be thought of as harbours for small, local populations of plants which could not otherwise persist in a habitat. Sites of disturbance are open systems because populations which establish in them are open to exchange of propagules with other sites. The conditions which determine whether coexistence between species is possible in open populations are different to those which apply in closed populations at equilibrium.

Consider a collection of gaps in a forest, a proportion of which each contain a few individual birch trees. If each of these gaps is gradually invaded by oaks, how long will it be before birch is no longer to be found anywhere in the forest? Obviously the exclusion of birch is only a matter of time unless new gaps are opened in the forest canopy, and these are seeded from gaps which already contain birch. Thus any individual birch population within a particular gap has a finite life expectancy but, so long as new gaps appear, the birch population of the forest as a whole may last indefinitely. Two things are important in this process:

1. The forest is not in a static condition, but contains disturbances which upset the equilibrium species composition locally prevent the universal dominance of the forest by one species such as oak.
2. Gaps are open, allowing initial invasion by birch and later dissemination of seeds.

An analogy can be drawn between a system of gap populations, interconnected by the migration of propagules, and an array of light bulbs wired together. An individual light bulb represents a gap which may be colonized (light *on*) or empty (light *off*). Just as gaps may be open and connected to each other by the migration of propagules, light bulbs can be wired so that their behaviour depends upon the state (*on* or *off*) of other bulbs in the array. We can also model what would happen in a system of closed gaps, where connections between gaps were broken, by breaking the connections between bulbs.

Now let us say that all bulbs in an unconnected array of 100 bulbs are *on*, and that each has a probability of 0.5 that it will go *off* in each second. This is equivalent to 100 gaps, all containing populations of birch which have an equal probability of extinction. For such an unconnected array it will be only about 2 seconds before all 100 lights have gone out. The extinction of 100 isolated gap populations of birch will eliminate birch from the forest entirely.

If gap populations are open and connected, an empty gap may be filled from an occupied one before this source of seeds also becomes extinct. At the extreme of connectedness in the forest of 100 gaps, we can envisage a situation where any gap which still contains birch may act as a source of seeds to any other gap which is empty. At this extreme all 100 bulbs in the array are connected to all the rest and are wired so that any bulb which is *off and* which is connected to at least 1 bulb that is *on* has a probability of being switched on of 0.5 in each second. Bulbs which are lit still have a probability of 0.5 in each second that they will turn *off*. In a connected array of this kind it is estimated that it would take 10^{22} years before all bulbs were turned *off*. The estimated age of the earth is only about 5×10^9 years.

A system of gap populations connected by migration clearly has great potential resilience against the chance event that many simultaneous local extinctions will wipe out the entire population at once. This explains how one species may survive by colonizing gaps, but what happens when two or three species play this game and all utilize the same gaps?

This situation has been modelled by Caswell (1978). Providing that gap-colonizing species are rare, so that encounters between different species in the same gap are relatively uncommon and providing that extinction caused by interference between species does not occur too rapidly, Caswell found that three species can coexist more or less indefinitely. This coexistence does not depend upon niche differences between species of the kind exhibited by *Lythrum* and *Epilobium*, but it does depend on the rate at which one species excludes another from a gap being far slower than was the case for these two species. In fact, it depends upon the outcome of such interspecific interference being delayed or protracted until sufficient new gaps have been created by disturbance. This allows a fresh start to be made in the struggle between coexisting species. The ultimate result of that struggle in any one place is local extinction for one or more species, but the chances of this producing global extinction of a species in an area are remote so long as disturbances continue to occur.

In most habitats there are various technical difficulties involved in measuring with the necessary accuracy how often gaps appear, when they do so and how long they last. Without such measurements, a kind of demography of gaps, it is not possible to tell whether the plants which colonize them when they are available may have sufficient resources to persist in the face of interference from the surrounding vegetation. It is no good being a gap colonizer if there are no gaps! Investigations of the role of disturbance in the ecology of populations are most advanced in marine habitats in the littoral and sub-littoral zones (e.g. Dayton 1971; Lieberman, John and Liberman 1979).

A study of the role of disturbance in the regeneration of the sea palm (*Postelsia palmaeformis*), a brown alga which occurs on the coast of Washington State, USA, provides a model which terrestrial ecologists could usefully emulate (Paine 1979). The alga grows attached to rock in gaps created by wave action in beds of the mussel *Mytilus californicus*. Mussels cover the rock surface, preventing attachment of *Postelsia* where no bare rock is present and encroaching on such gaps as there are, gradually closing them to colonization by the plant. *Postelsia* is an annual and must re-establish every year in order to persist at a particular site. Paine collected data on the area of bare rock surface exposed annually by wave action at 26 sites for 11 years. Sites with low disturbance (less than 3% of rock surface bared each year) carried no population of *Postelsia* but about one-third of the sites with greater disturbance did have a population of the alga. On comparing the high-disturbance sites with and without *Postelsia*, Paine found that both had the same mean rate of gap formation (about 7% of surface area per year), but that these mean rates were the result of disturbances created in significantly different ways. At the sites with *Postelsia* present, disturbance was a regular annual event between November and April with a relatively low variation in the area of rock surface exposed each year. At the sites where *Postelsia* was absent, disturbance by wave action also occurred annually between November and April but the area of rock surface exposed in different years was extremely variable. In some years little rock would be bared and in others much more.

Paine concluded that *Postelsia* was excluded from the sites of low disturbance by the lack of bare rock for colonization and from the sites of very variable disturbance by the extinction of populations caught in long intervals when no new gaps were available. The latter hypothesis was tested by establishing *Postelsia* phytometers at sites of high mean disturbance where they were normally absent. Some regeneration of phytometer populations occurred at these sites but it was insufficient to maintain them and plant density declined as mussels filled available gaps. Populations occurring at a density of less than 25 m^{-2} were unable to replace themselves at all.

Summary

The resources a population requires to maintain population size and its manner of exploiting these describe its *niche*. The *competitive exclusion principle* states that two species will not coexist in a stable equilibrium mixture if they share the same limiting factor.

Two species may coexist when each inhibits its own population growth more than that of its competitor. *Frequency-dependent interference* is an example of this situation. Species which partition a common limiting resource are described collectively as a *guild*.

The concept of the niche may be resolved into two components: the *fundamental niche* and the *realized niche*. *Phytometer* experiments can be used to map out a species' niche and to determine the effects of interference from other species. Phytometers may take a long time to succumb to an alien environment. The regeneration phase is the most vulnerable in a phytometer population.

In a stable environment species may avoid interference through niche differences. In an unstable environment, disturbance may prevent competitive exclusion taking place. The competitive exclusion principle is not applicable in open systems subject to recolonization from outside. The nature and frequency of disturbance may determine whether a colonizing species is able to persist in the face of interference from other species or not.

Bibliography

Abrahamson, W.G. (1980) Demography and vegetative reproduction, Ch. 5, pp. 89–106, in Solbrig, O.T. (ed.) *Demography and evolution in plant populations*. Blackwell, Oxford; University of California Press, California.

Abrahamson, W.G. and Gadgil, M. (1973) Growth form and reproductive effort in goldenrods (*Solidago*, compositae). *Am. Nat.*, **107**, 651–61.

Allen, E.B. and Forman, R.T.T. (1976) Plant species removals and old-field community structure and stability, *Ecology*, **57**, 1233–43.

Arthur, A.E. Gale, J.S. and Lawrence, K.J. (1973) Variation in wild populations of *Papaver dubium*: VII. Germination time, *Heredity*, **30**, 189–97.

Attsat, P.R. and O'Dowd, D.J. (1976) Plant defence guilds. *Science*, **193**, 24–9.

Auclair, A.N. and Cottam, G. (1971) Dynamics of black cherry (*Prunus serotina* Erhr.) in Southern Wisconsin oakforests, *Ecol. Monogr.*, **41**, 153–77.

Bach, C.E. (1980) Effects of plant diversity and time of colonization on a herbivore–plant interaction, *Oecologia*, **44**, 319–26.

Baker, H.G. (1972) Seed weight in relation to environmental conditions in California, *Ecology*, **53**, 997–1010.

Barkham, J.P. (1980) Population dynamics of the wild daffodil (*Narcissus pseudonarcissus*). In clonal growth, seed reproduction, mortality and the effects of density, *J. Ecol.*, **68**, 607–33.

Barnes, B.V. (1966) The clonal growth habit of American aspens, *Ecology*, **47**, 439–47.

Baskin, J.M. and Baskin, C.C. (1972) Influence of germination date on survival and seed production in a natural population of *Leavenworthia stylosa*, *Am. Midl. Nat.*, **88**, 318–23.

Baskin, J.M. and Baskin, C.C. (1974) Germination and survival in a population of the winter annual *Alyssum alyssoides*, *Can. J. Bot.*, **52**, 2439–45.

Baskin, J.M. and Baskin, C.C. (1979a) Studies on the autecology and population biology of the monocarpic perennial *Grindelia lanceolata*, *Am. Midl. Nat.*, **41**, 290–99

Baskin, J.M. and Baskin, C.C. (1979b) Studies on the autecology and population biology of the weedy monocarpic perennial, *Pastinaca sativa*, *J. Ecol.*, **67**, 601–10.

Bazzaz, F.A. and Carlson, R.W. (1979) Photosynthetic contribution of flowers and seeds to reproductive effort of an annual colonizer, *New Phytol.*, **82**, 223–32.

Bazzaz, F.A. and Harper, J.L. (1976) Relationship between plant weight and numbers in mixed populations of *Sinapis alba* (L) Rabenh and *Lepidium sativum* L, *J. Appl. Ecol.*, **13**, 211–16.

Bell, A.D. (1974) Rhizome organization in relation to vegetative spread in *Medeola virginiana*, *J. Arnold Arboretum*, **55**, 458–68.

Bell, A.D. (1979) The hexagonal branching pattern of *Alpinia speciosa* L. (Zingiberaceae), *Ann. Bot.*, **43**, 209–23.

Bell, A.D. and Tomlinson, P.B. (1980) Adaptive architecture in rhizomatous plants, *Bot. J. Linn. Soc.*, **80**, 125–60.

Bergh, J.P. Van den, and Wit, de (1960) Concurrentie tussen thimothee (*Phleum pratense* L.) en reukgras (*Anthoxanthum odoratum* L.), *Jaarboek I.B.S.*, 155–66.

Bishop, G.F. Davy A.J. and Jeffries, R.L. (1978) Demography of *Hieracium pilosella* in a Breck grassland, *J. Ecol.*, **66**, 615–29.

Black, J.N. (1958) Competition between plants of different initial seed sizes in swards of subterranean clover (*Trifolium subterraneum* L.) with particular reference to leaf area and the light microclimate, *Aust. J. Agric. Res.*, **9**, 299–318.

Blackman, G.E. and Rutter, A.J. (1950) Physiological and ecological studies in the analysis of plant environment: V. An assessment of the factors controlling the distribution of the bluebell (*Scilla nonscripta*) in different communities, *Ann. Bot.*, **14**, 487–520.

Borchert, M.I. and Jain, S.K. (1978) The effect of rodent seed predation on four species of California annual grasses, *Oecologia*, **33**, 101–13.

Bossema, I. (1979) Jays and oaks: an eco-ethological study of a symbiosis, *Behaviour*, **70**, 1–117.

Bourdeau, P. (1954) Oak seedling ecology determining segregation of species in Piedmont oak–hickory forest, *Ecol. Monogr.*, **24**, 297–320.

Bradshaw, A.D. (1965) Evolutionary significance of phenotypic plasticity in plants, *Adv. Genet.*, **13**, 115–55.

Bradshaw, M.E. and Doody, J.P. (1978a) Plant population studies and their relevance to nature conservation, *Biol. Conserv.*, **14**, 223–42.

Bradshaw, M.E. and Doody, J.P. (1978b) Population-dynamics and biology, Ch. 2. pp. 48–63 in Chapham, A.R. (ed.), *Upper Teesdale, the area and its natural history.* Collins, London.

Brenchley, W.E. and Warrington, K. (1930) The weed seed population of arable soil: I. Numerical estimation of viable seeds and observations on their natural dormancy, *J. Ecol.*, **18**, 235–72.

Brockway, L.H. (1980) *Science and colonial expansion: the role of the British Royal Botanic Gardens.* Academic Press, London and New York.

Brown, J.H. Reichman, O.J. and Davidson, D.W. (1979) Granivory in desert ecosystems, *Ann. Rev. Ecol. Syst.*, **10**, 201–27

Buchanan, G.A. Crowley, R.H. Street, J.E. and McGuire, J.A. (1980) Competition of sicklepod (*Cassia obtusifolia*) and redroot pigweed (*Amaranthus retroflexus*) with cotton (*Gossypium hirsutum*), *Weed Sci.*, **28**, 258–62.

Budd, A.C. Chepil, W.S. and Doughty, J.L. (1954) Germination of weed seeds: II. The influence of crops and fallow on the weed seed population of the soil, *Can. J. Agric. Sci.*, **34**, 18–27.

Burdon, J.J. and Chilvers, G.A. (1975) Epidemiology of damping-off disease (*Pythium irregulare*) in relation to density of *Lepidium sativum* seedlings, *Ann. Appl. Biol.*, **81**, 135–43.

Callaghan, T.V. (1976) Strategies of growth and population dynamics of plants: 3. Growth and population dynamics of *Carex bigelowii* in an alpine environment, *Oikos*, **27**, 402–13.

Canfield, R.H. (1957) Reproduction and life span of some perennial grasses of southern Arizona, *J. Range Mgmt*, **10**, 199–203.

Caswell, H. (1978) Predator mediated coexistence: a non-equilibrium model, *Am. Nat.*, **112**, 127–54.

Champness, S.S. and Morris, K. (1948) The population of buried weed seeds in relation to contrasting pasture and soil types. *J. Ecol.*, **36**, 149–73.

Chancellor, R.J. (1968) The value of biological studies in weed control, *Proc. 9th Br. Weed Control Conf.*, 1129–35.

Chew, R.M. and Chew, A.E. (1965) The primary productivity of a desert-shrub (*Larrea tridentata*) community, *Ecol. Monogr.*, **35**, 355–75.

Chilvers, G.A. and Brittain, E.G. (1972) Plant competition mediated by host-specific parasites – a simple model, *Aust. J. Biol. Sci.*, **25**, 748–56.

Chippendale, H.G. and Milton, W.E.J. (1934) On the viable seeds present in the soil beneath pastures, *J. Ecol.*, **22**, 508–31.

Clements, F.E. Weaver, J.E. and Hanson, H.C. (1929) *Plant competition, Carnegie, Inst. Wash. Pub.*, p. 398.

Clymo, R.S. and Reddaway, E.J.F. (1972) A tentative dry matter balance sheet for the wet blanket bog on Burnt Hill, Moor House NNR, *Moor House Occasional Paper*, No 3, Nature Conservancy, London.

Cody, M.L. (1966) A general theory of clutch size, *Evolution*, **20**, 174–84.

Cole, L.C. (1954) The population consequences of life history phenomena, *Quart. Rev. Biol.*, **29**, 103–37.

Collins, N.J. (1976) Growth and population dynamics of the moss *Polytrichum alpestre* in the maritime Antarctic. Strategies of growth and population dynamics of tundra plants, 2, *Oikos*, **27**, 389–401.

Cook, R.E. (1980) Germination and size-dependent mortality in *Viola blanda*, *Oecologia*, **47**, 115–17.

Courtney, A.D. (1968) Seed dormancy and field emergence in *Polygonum aviculare*, *J. Appl. Ecol.*, **5**, 675–84.

Crawford-Sidebotham, T.J. (1972) The role of slugs and snails in the maintenance of the cyanogenesis polymorphism of *Lotus corniculatus* L. and *Trifolium repens* L., *Heredity*, **28**, 405–11.

Crisp. M.D. and Lange, R.T. (1976) Age structure distribution and survival under grazing of the arid zone shrub *Acacia burkitti*, *Oikos*, **27**, 86–92.

Crow, T.R. (1980) A rainforest chronicle – a 30-year record of change in structure and composition at El-Verde, Puerto-Rico, *Biotropica*, **12**, 42–55.

Culver, D.C. and Beatie, A.J. (1978) Myrmecochory in viola: dynamics of seed-ant interactions in some west Virginian species, *J. Ecol.*, **66**, 53–72.

Curtis, J.T (1959) *The vegetation of Wisconsin*. University of Wisconsin Press, Madison, Wisconsin.

Darwin, C. (1859) *The origin of species*. 1st edn, Murray, London;

Dayton, P.K. (1971) Competition, disturbance and community organization: the provision and subsequent utilization of space in a rocky intertidal community, *Ecol. Monogr.*, **41**, 351–89.

Deevey, E.S. (1947) Life tables for natural populations of animals, *Quart. Rev. Biol.*, **22**, 283–314.

Dirzo, R and Harper, J.L. (1980) Experimental studies on slug–plant interactions: II. The effect of grazing by slugs on high density monocultures of *Capsella bursa-pastoris* and *Poa annua*, *J. Ecol.*, **68**, 999–1011.

Downs, A.A. (1944) Estimating acorn crops for wild life in the southern Appalachians, *J. Wildl. Mgmt*, **8**, 339–40.

Downs, C. and McQuilkin, W.E (1944) Seed production of Southern Appalachian oaks. *Journal of Forestry*, **42**, 913–20.

Duckett, J.G. and Duckett, A.R. (1980) Reproductive biology and population dynamics of wild gametophytes of *Equisetum,*, *Bot. J. Linn. Soc.*, **80**, 1–40.

Eis, S. Garman, E.H. and Ebel, L.F. (1965) Relation between cone production and diameter increment of douglas fir (*Pseudotsuga menziesii* (Mirb.) Franco), grand fir (*Abies grandis* Dougl.), and western white pine (*Pinus monticola* Dougl.), *Can. J. Bot.*, **43**, 1553–9.

Ennos, R.A. (1981) Detection of selection in populations of white clover (*Trifolium repens* L.), *Biol. J. Linn. Soc.*, **15**, 75–82.

Epling, C. Lewis, H. and Ball, E.M. (1960) The breeding group and seed storage: a study in population dynamics, *Evolution*, **14**, 238–55.

Fagerström, T. and Agren, G.I. (1979) Theory for coexistence of species differing in regeneration properties, *Oikos*, **33**, 1–10.

Fisher, R.A. (1930) *The genetical theory of natural selection*. Oxford University Press.

Flor, H.H. (1956) The complimentary genic systems in flax and flax rust, *Adv. Genet.*, **8**, 29–54.

Flor, H.H. (1971) Current status of the gene-for-gene concept, *Ann. Rev. Phytopathol.*, **9**, 275–96

Flower-Ellis, J.G.K. (1971) Age structure and dynamics in stands of bilberry, *Rapp. Uppsatts. Avdel. Skogsekol*, **9**, 1–108.

Ford, E.D. (1975) Competition and stand structure in some even-aged plant monocultures, *J. Ecol.*, **63**, 311–33.

Fortainier, E.J. (1973) Reviewing the length of the generation period and its shortening, particularly in tulips, *Sci. Hort.*, **1**, 107–16.

Fowells, H.A. and Schubert, G.H. (1956) Seeds crops of forest trees in the pine region of California. *USDA Tech. Bull.*, **1150**, 48pp.

Fowler, N.L. (1981) Competition and coexistence in a North Carolina grassland: II. The effects of the experimental removal of species, *J. Ecol.*, **69**, 825–41.

Fridrikson, S. (1975) *Surtsey, evolution of life on a volcanic island*, Butterworth, London.

Frissell, S.S. (1973) The importance of fire as a natural ecological factor in Itasca State Park, Minnesota, *Quat. Res.*, **3**, 397–407.

Gadgil P.M. Solbrig, O.T. (1972) The concept of *r* and *K* selection. Evidence from some wild flowers and theoretical considerations, *Am. Nat.*, **106**, 14–31.

Gashwiler, J.S. (1967) Conifer seed survival in a western Oregon clearcut, *Ecology*, **48**, 431–3.

Gause, G.F. (1934) *The struggle for existence*. Williams and Wilkins, Baltimore.

Gleason, H.A. (1926) The individualistic concept of the plant association, *Bull. Torrey Bot. Club*, **53**, 7–26.

Gleason, H.A. (1927) Further views on the succession-concept, *Ecology*, **8**, 299–326.

Glier, J.H. and Caruso, J.L. (1973) Low-temperature induction of starch degradation in roots of a biennial weed, *Cryobiology*, **10**, 328–30.

Golubeva, J.N. (1962) Some data on pools of viable seeds in soil under meadow-steppe vegetation (In Russian), *Byull. Mosk. Obshch. Isp. Prir.*, **67**, 76–89.

Gould, S.J. and Lewontin, R.C. (1979) The spandrels of San Marco and the Panglossian paradigm: a critique of the adaptationist programme, *Proc. R. Soc. Lond. B.*, **205**, 581–98.

Grant, M.C. and Antonovics, J. (1978) Biology of ecologically marginal populations of *Anthoxanthum odoratum*: I. Phenetics and dynamics, *Evolution*, **32**, 822–38.

Gray, B. (1972) Economic tropical forest entomology, *Ann. Rev. Entomol.*, **17**, 313–54.

Gray, B. (1975) Size-composition and regeneration of *Araucaria* stands in *New Guinea J. Ecol.*, **63**, 273–89.

Grime, J.P. (1979) *Plant strategies and vegetation processes*. Wiley, Chichester, UK and New York, USA.

Grime, J.P. and Jeffrey, D.W. (1965) Seedling establishment in vertical gradients of sunlight, *J. Ecol.*, **53**, 621–42.

Gross, K.L. (1980) Colonization by *Verbascum thapsus* (Mullein) of an old field in Michigan: experiments on the effects of vegetation, *J. Ecol.*, **68**, 919–28.

Groves, R.H. and Williams, J.D. (1975) Growth of skeleton weed (*Chondrilla juncea* L.) as affected by growth of subterranean clover (*Trifolium subterraneum* L.) and infection by *Puccinia chondrilla* Bubak and Syd, *Aust. J. Agric. Res.*, **26**, 975–83.

Grubb, P.J. (1976) A theoretical background to the conservation of ecologically distinct groups of annuals and biennials in the chalk grassland ecosystem, *Biol. Conserv.*, **10**, 53–76.

Grubb, P.J. (1977) The maintenance of species richness in plant communities. The importance of the regeneration niche, *Biol. Rev.*, **52**, 107–45.

Guittet, J. and Laberche, J.C. (1974) L'implantation naturelle du pin sylvestre sur pelouse xérophile en foret de Fontainebleau: II. Demographie des graines et des plantules au voisinage des vieux arbres, *Oecol. Plant.*, **9**, 111–30.

Gunnill, F.C. (1980) Demography of the intertidal brown alga *Pelvetia fastigiata* in Southern California, USA *Marine Biol.*, **59**, 169–79.

Hairston, N.G. Tinkle, D.W. and Wilbur, H.M. (1970) Natural selection and the parameters of population growth, *J. Wildl. Mgmt*, **34**, 681–90.

Haizel, K.A. and Harper, J.L. (1973) The effects of density and the timing of removal on interference between barley, white mustard and wild oats, *J. Appl. Ecol.*, **10**, 23–32.

Hall, R.L. (1974) Analysis of the nature of interference between plants of different species: II. Nutrient relations in a Nandi *Setaria* and greenleaf *Desmodium* association with particular reference to potassium, *Aust. J. Agric. Res.*, **25**, 749–56.

Hanf, M. (1974) *Weeds and their seedlings*. BASF, Ipswich, England.

Harberd, D.J. (1961) Observations on population structure and longevity of *Festuca rubra* L., *New Phytol.*, **60**, 184–206.

Harberd, D.J. (1962) Some observations on natural clones in *Festuca ovina*, *New Phytol.*, **61**, 85–100.

Harberd, D.J. (1963) Observations on natural clones of *Trifolium repens* L., *New Phytol.*, **62**, 198–204.

Harcourt, D.G. (1970) Crop life tables as a pest management tool, *Can. Entomol.*, **102**, 950–5.

Harper, J.L. (1959) The ecological significance of dormancy and its importance in weed control, Proc. 4th Int. Congr. Crop. Prot. (*Hamburg*), 415–20.

Harper, J.L. (1961) Approaches to the study of plant competition, *Soc. Exp. Biol. Symp.*, **15**, 1–39.

Harper, J.L. (1977) *Population biology of plants*. Academic Press, London and New York.

Harper, J.L. (1978) The demography of plants with clonal growth, pp. 27–48 in Freysen, A.H.J. and Woldendorp, J. (eds), *Structure and functioning of plant populations*. North Holland Publ. Co., Amsterdam.

Harper, J.L. and Gajic, D. (1961) Experimental studies of the mortality and plasticity of a weed, *Weed Res.*, **1**, 91–104.

Harper, J.L. Lovell, P.H. and Moore, K.G. (1970) The shapes and sizes of seeds, *Ann. Rev. Ecol. Syst.*, **1**, 327–56.

Harper, J.L. and Ogden, J. (1970) The reproductive strategy of higher plants: I. The concept of strategy with special reference to *Senecio vulgaris* L., *J. Ecol.*, **58**, 681–98.

Harper, J.L. and McNaughton, I.H. (1962) The comparative biology of closely related species living in the same area: VII. Interference between individuals in pure and mixed populations of *Papaver* spp., *New. Phytol.*, **61**, 175–88.

Harper, J.L. and White, J. (1971) The dynamics of plant populations, *Proc. Adv. Study Inst. Dynamics Numbers Popul.* (*Oosterbeek 1970*), 41–63.

Harper, J.L. Williams, J.T. and Sagar, G.R. (1965) The behaviour of seeds in the soil: I. The heterogeneity of soil surfaces and its role in determining the establishment of plants from seed, *J. Ecol.*, **53**, 273–86.

Hart, R. (1977) Why are biennials so few? *Am. Nat.*, **111**, 792–9.

Hartshorn, G.S. (1977) Tree falls and tropical forest dynamics, pp. 617–38 in Tomlinson, P.B. and Zimmerman, M.H. (eds), *Tropical trees as living systems*. Cambridge University Press, Cambridge and New York.

Haslam, S.M. (1978) *River plants*. Cambridge University Press, Cambridge and New York.

Hawthorne, W. and Cavers, P.B. (1976) Population dynamics of the perennial herbs *Plantago major* L. and *P. rugeli Decne*, *J. Ecol.*, **64**, 511–27.

Hayashi, I. and Numata, M. (1971) Ecological studies on the buried seed population in the soil related to plant succession: IV. *Jap. J. Ecol.*, **20**, 243–52.

Heinselman, M.L. (1973) Fire in the virgin forests of the Boundary Waters Canoe Area, Minnesota, *Quat. Res.*, **3**, 329–82.

Hett, J.M. (1971) A dynamic analysis of age in sugar maple seedlings, *Ecology*, **52**, 1071–4.

Hett, J.M. and Loucks, O.L. (1971) Sugar maple (*Acer saccharum* Marsh.) seedling mortality, *J. Ecol.*, **59**, 507–20.

Hett, J.M. and Loucks, O.L. (1976) Age structure models of balsam fir and eastern hemlock, *J. Ecol.*, **64**, 1029–44.

Hibbs, D.E. (1979) The age structure of a striped maple population, *Can. J. For. Res.*, **9**, 504–8.

Hibbs, D.E. and Fischer, B.C. (1979) Sexual and vegetative reproduction of striped maple (*Acer pensylvanicum* L.), *Bull. Torrey Bot. Club*, **106**, 222–27.

Hickman, J.C. (1975) Environmental unpredictability and plastic energy allocation strategies in the annual *Polygonum cascadense* (Polygonaceae), *J. Ecol.*, **63**, 689–701.

Hiroi, T. and Monsi, M. (1966) Dry-matter economy of *Helianthus annuus communities grown at varying densities and light intensities*, *J. Fac. Sci. Tokyo Univ.*, **III-9**, 241–85.

Holm, T. (1899) *Podophyllum peltatum*, a morphological study, *Bot. Gaz.*, **27**, 419–43.

Holmsgaard, E (1956) Effect of seed-bearing on the increment of European beech (*Fagus sylvatica* L.) and Norway spruce (*Picea abies* (L) Karst), *Proc. Int. Univ. For. Res. Org.*, *12th Congr.*, Oxford, 158–61.

Holt, B.R. (1972) Effect of arrival time on recruitment mortality and reproduction in successional plant populations, *Ecology*, **53**, 668–73.

Horn, H.S. (1966) Measurement of 'overlap' in comparative ecological studies, *Am. Nat.*, **100**, 419–24.

Hubbell, S.P. (1980) Seed predation and coexistence of tree species in tropical forests. *Oikos*, **35**, 214–29.

Huston, M (1979) A general hypothesis of species diversity, *Am. Nat.*, **113**, 81–101.

Hutchings, M.J, (1979) Weight–density relationships in ramet populations of clonal perennial herbs, with special reference to the $-3/2$ power law, *J. Ecol.*, **67**, 21–33.

Hutchinson, G.E. (1957) The multivariate niche, *Cold. Spr. Harb. Symp. Quant. Biol.*, **22**, 415–21.

Ikusima, I. and Shinozaki, K. (1955) Intraspecific competition among higher plants: II. Growth of duckweed, with a theoretical consideration on the C–D effect, *J. Inst. Polytech. Osaka. Univ.*, Ser. D6, 107–19.

Janzen, D.H. (1969) Seed-eaters versus seed size, number, toxicity and dispersal, *Evolution*, **23**, 1–27.

Janzen, D.H. (1970) Herbivores and the number of tree species in tropical forests, *Am. Nat.*, **104**, 501–28.

Janzen, D.H. (1971) Escape of *Casia grandis* L. beans from predators in time and space, *Ecology*, **52**, 964–79.

Janzen, D.H. (1972) Escape in space by *Sterculia apetala* seeds from the bug *Dysdercus fasciatus* in a Costa Rican deciduous forest, *Ecology*, **53**, 350–61.

Janzen, D.H. (1973) Dissolution of mutualism between *Ceropia* and its *Azteca* ants, *Biotropica*, **5**, 15–28.

Janzen, D.H. (1975a) *Ecology of plants in the tropics*. Studies in Biology No. 58, Arnold, London.

Janzen, D.H. (1975b) Behaviour of *Hymenaea coubaril* when its predispersal seed predator is absent, *Science*, **189**, 145–7.

Janzen, D.H. (1976) Why bamboos wait so long to flower, *Ann. Rev. Ecol. Syst.*, **7**, 347–91.

Janzen, D.H. (1980) Specificity of seed-attacking beetles in a Costa Rican deciduous forest, *J. Ecol.*, **68**, 929–52.

Jeffries, R.L. Davy, A.J. and Rudmik, T. (1981) Population biology of the salt marsh annual *Salicornia europaea* agg., *J. Ecol.*, **69**, 17–31.

Johnson, E.A. (1975) Buried seed populations in the subarctic forest east of Great Slave Lake, Northwest Territories, *Can. J. Bot.*, **53**, 2933–41.

Johnson, M.P. and Cook, S.A. (1968) 'Clutch size' in buttercups, *Am. Nat.*, **102**, 405–11.

Jong, de, P. Aarssen, L.W. and Turkington, R. (1980) The analysis of contact sampling data, *Oecologia*, **45**, 322–4.

Karpov, V.G. (1960) On the species composition of the viable seed supply in the soil of spruce *Vaccinium myrtillus* vegetation (in Russian), *Trudy Mosk. Obshch. Isp. Priorody*, **3**, 131–40.

Kawano, S. (1975) The productive and reproductive biology of flowering plants: II. The concept of life history strategy in plants, *J. Coll. Lib. Arts Toyama Univ. Japan*, **8**, 51–86.

Kawano, S. and Masuda, J. (1980) The productive and reproductive biology of flowering plants: VII. Resource allocation and reproductive capacity in wild populations of *Heloniopsis orientalis* (Thunb.) C. Tanaka (Liliaceae), *Oecologia*, **45**, 307–17.

Kawano, S. and Nagai, Y. (1975) The productive and reproductive biology of flowering plants: I. Life history strategies of three *Allium* species in Japan, *Bot. Mag. Tokyo*, **88**, 281–318.

Kays, S. and Harper, J.L. (1974) The regulation of plant and tiller density in a grass sward, *J. Ecol.*, **62**, 97–105.

Kellman, M.C. (1970) The viable seed content of some forest soils in coastal British Columbia, *Can. J. Bot.*, **48**, 1383–5.

Kellman, M.C. (1974a) The viable weed seed content of some tropical agricultural soils, *J. Appl. Ecol.*, **11**, 669–77.

Kellman, M.C. (1974b) Preliminary seed budgets for two plant communities in coastal British Columbia, *J. Biogeography*, **1**, 123–33.

Klemow, K.M. and Raynal, D.J. (1981) Population ecology of *Melilotus alba* in a limestone quarry, *J. Ecol.*, **69**, 33–44.

Kulman, H.M. (1971) Effects of insect defoliation on growth and mortality of trees, *Ann. Rev. Ent.*, **16**, 289–324.

Lack, D. (1954) *Natural regulation of animal numbers*. Clarendon Press, Oxford.

Laine, K.J. and Niemela, P. (1980) The influence of ants on the survival of mountain birches during an *Oporinia autumnata* (Lep., Geometridae) outbreak, *Oecologia*, **47**, 39–42.

Langer, R.H.M. Ryle, S.M. and Jewiss, O.R. (1964) The changing plant and tiller populations of timothy and meadow fescue swards: I. Plant survival and the pattern of tillering, *J. Appl. Ecol.*, **1**, 197–208.

Larson, M.M. and Shubert, G.H. (1970) Cone crops of ponderosa pine in central Arizona, including the influence of Abert squirrels. *USDA Forest Serv. Res. Pap.* R.M., **58**, 15pp.

Law, R. (1979) The cost of reproduction in annual meadow grass, *Am. Nat.*, **113**, 3–16.

Law, R: Bradshaw, A.D. and Putwain, P.D. (1977) Life history variation in *Poa annua*, *Evolution*, **31**, 233–46.

Leak, W.B. (1975) Age distributions in virgin red spruce and northern hardwoods, *Ecology*, **56**, 1451–4.

Leck, M.A. and Graveline, K.J. (1979) The seed bank of a freshwater tidal marsh, *Am. J. Bot.*, **66**, 1006–15.

Leverich, W.J. and Levin, D.A. (1979) Age-specific survivorship and reproduction in *Phlox drummondii*, *Am. Nat.*, **113**, 881–903.

Levin, D.A. and Kerster, H.W. (1974) Gene flow in seed plants, *Evol. Biol.*, **7**, 139–220.

Levin, D.A. and Turner, B.L. (1977) Clutch size in the compositae, Ch. 18, pp. 215–22 in Stonehouse, B. and Perrins, C.M. (eds), *Evolutionary ecology*. Macmillan, London.

Lieberman, M. John, D.M. and Liberman, D. (1979) Ecology of subtidal algae on seasonally devastated cobble substrates off Ghana, *Ecology*, **60**, 1151–61.

Lieth, H. (1960) Patterns of change within grassland communities, pp. 27–39 in Harper, J.L. (ed.), *The biology of weeds*. Blackwell, Oxford.

Ligon, D.J. (1978) Reproductive interdependence of pinyon jays and pinyon pines, *Ecol. Monogr.*, **48**, 111–26.

Lippert, R.D. and Hopkins, H.H. (1950) Study of viable seeds in various habitats in mixed prairies, *Trans. Kansas Acad. Sci.*, **53**, 355–64.

Lloyd, D.G. (1981) Sexual strategies in plants: I. An hypothesis of serial adjustment of maternal investment during one reproductive session, *New Phytol.*, **86**, 69–79.

Lovett Doust, J. (1980) Experimental manipulation of patterns of resource allocation in the growth cycle and reproduction of *Smyrnium olusatrum* L. *Biol. J. Linn. Soc.*, **13**, 155–66.

MacArthur, R.H. (1972) *Geographical ecology*. Harper Row, New York.

MacArthur, R.H. and Levins, R. (1967) The limiting similarity, convergence and divergence of coexisting species, *Am. Nat.*, **101**, 377–85.

MacArthur, R.H. and Wilson, E.O. (1967) *The theory of island biogeography*. Princeton University Press, Princeton, N.J.

Mack, R.N. (1976) Survivorship of *Cerastium atrovirens* at Aberffraw, Anglesey, *J. Ecol.*, **64**, 309–12.

Mack, R.N. and Harper, J.L. (1977) Interference in dune annuals: spatial pattern and neighborhood effects, *J. Ecol.*, **65**, 345–64.

Major, J. and Pyott, W.T. (1966) Buried viable seeds in California bunchgrass sites and their bearing on the definition of a flora, *Vegetatio Acta Geobotanica*, **13**, 253–82.

Marchand, P.J. and Roach, D.A. (1980) Reproductive strategies of pioneering alpine species: seed production, dispersal, and germination, *Arc. Alp. Res.*, **12**, 137–46.

Marks, M. and Prince, S. (1981) Influence of germination date on survival and fecundity in wild lettuce *Lactuca serrriola, Oikos*, **36**, 326–30.

Marks, P.L. (1974) The role of pin cherry (*Prunus pennsylvanica* L.) in the maintenance of stability in northern hardwood ecosystems, *Ecol. Monogr.*, **44**, 73–88.

May, R.M. (1975) Some notes on estimating the competition matrix α, *Ecology*, **56**, 737–41.

Mayer, A.M. and Polijakof-Mayber, A. (1975) *The germination of seeds*. Pergamon, Oxford.

Mellanby, K. (1968) The effects of some mammals and birds on regeneration of oak, *J. Appl. Ecol.*, **5**, 359–66.

Milton, W.E.J. (1939) The occurrence of buried viable seeds in soils at different elevations and on a salt marsh, *J. Ecol.*, **27**, 149–59.

Mohler, C.L. Marks, P.L. and Sprugel, D.G. (1978) Stand structure and allometry of trees during self-thinning of pure stands, *J. Ecol.*, **66**, 599–614.

Murphy, G.I. (1968) Pattern in life history and the environment, *Am. Nat.*, **102**, 390–404.

Naylor, R.E.L. (1972) Aspects of the population dynamics of the weed *Alopecurus myosuroides* Huds. in winter cereal crops, *J. Appl. Ecol.*, **9**, 127–39.

Nelson, J.F. and Chew, R.M. (1977) Factors affecting seed reserves in the soil of a Mojave desert ecosystem, Rock valley, Nye County, Nevada, *Am. Midl. Nat.*, **97**, 300–20.

Niemela, P. Tuomi, J. and Haukioja, E. (1980) Age-specific resistance in trees: defoliation of tamaracks (*Larix larcina*) by larch bud moth (*Zeiraphera improbana*) (Lep., Tortricidae), *Rep. Kevo Subarctic Res. Stat.*, **16**, 49–57.

Ng, F.S.P. (1977) Strategies of establishment in Malayan forest trees, Ch. 5, pp. 129–62 in Tomlinson, P.B. and Zimmerman, M.H. (eds), *Tropical trees as living systems*. Cambridge University Press, Cambridge and New York.

Noble, J.C. Bell, A.D. and Harper, J.L. (1979) The population biology of plants with clonal growth: I. The morphology and structural demography of *Carex arenaria*, *J. Ecol.*, **67**, 983–1008.

O'Dowd, D.J. and Hay, M.E. (1980) Mutualism between harvester ants and a desert ephemeral: seed escape from rodents, *Ecology*, **61**, 531–40.

Odum, E.P. (1971) *Fundamentals of ecology*. 3rd edn, W.B. Saunders and Co., Philadelphia.

Odum, S. (1978) *Dormant seeds in Danish ruderal soils*. Horsholm Arboretum, Denmark.

Ogden, J. (1974) The reproductive strategy of higher plants: II. The reproductive strategy of *Tussilago farfara* L., *J. Ecol.*, **62**, 291–324.

Oinonen, E. (1967a) Sporal regeneration of ground pine (*Lycopodium complanatum* L.) in southern Finland in the light of the size and age of its clones, *Acta For. Fenn.*, **83**, 76–85.

Oinonen, E. (1967b) The correlation between the size of Finnish bracken (*Pteridium aquilinum* (L.) Kuhn) clones and certain periods of site history, *Acta For. Fenn.*, **83**, 1–51.

Oinonen, E. (1969) The time-table of vegetative spreading of the lily-of-the-valley (*Convallaria majalis* L.) and the wood small-reed (*Calamagrostis epigeios* (L.) Roth) in southern Finland, *Acta For. Fenn.*, **97**, 1–35.

Oomes, M.J. and Elberse, W.Th. (1976) Germination of six grassland herbs in microsites with different water contents, *J. Ecol.*, **64**, 743–55.

Oosting, H.J. and Humphries, M.E. (1940) Buried viable seed in a successional series of old field and forest soils, *Bull. Torrey Bot. Club*, **67**, 253–73.

Open University (1981) *Evolutionary Ecology*. Unit 11 of S364 Evolution: Science, a third level course. Open University Press, Milton Keynes.

Paine, R.T. (1979) Disaster catastrophe and local persistence of the sea palm *Postelsia palmaeformis*, *Science*, **205**, 685–7.

Palmblad, I.G. (1968) Competition studies on experimental populations of weeds with emphasis on the regulation of population size, *Ecology*, **49**, 26–34.

Perkins, D.F. (1968) Ecology of *Nardus stricta* L: I. Annual growth in relation to tiller phenology, *J. Ecol.*, **56**, 633–46.

Phillips, J. (1934) Succession development and climax and the complex organism: an analysis of concepts: Part I. *J. Ecol.*, **22**, 554–71.

Pinder, III. J.E. (1975) Effects of species removal on an old-field plant community, Ecology, **56**, 747–51.

Platt, W.J. (1975) The colonization and formation of equilibrium plant species associations on badger disturbances in a tall-grass prairie, *Ecol. Monogr.*, **45**, 285–305.

Platt, W.J. and Weiss, I.M. (1977) Resource partitioning and competition within a guild of fugitive prairie plants, *Am. Nat.*, **111**, 479–513.

Primack, R.B. (1979) Reproductive effort in annual and perennial species of *Plantago* (Plantaginaceae), *Am. Nat.*, **114**, 51–62.

Puckridge, D.W. and Donald, C.M. (1967) Competition among wheat plants sown at a wide range of densities, *Aust. J. Agric. Res.*, **17**, 193–211.

Purseglove, J.W. (1968) *Tropical crops. Dicotyledons*. Longman, London.

Rabotnov, T.A. (1964) The biology of monocarp perennial meadow plants, *Bull. Moscow Soc. Nature*, **69**, 57–66. Russian Translation Service 8739, British Library.

Rabotnov, T.A. (1978a) Structure and method of studying coenotic populations of perennial herbaceous plants, *Sov. J. Ecol.*, **9**, 99–105.

Rabotnov, T.A. (1978b) On coenopopulations of plants reproducing by seeds, pp. 1–26 in Freysen, A.H.J. and Woldendrop, J. (eds), *Structure and functioning of plant populations*. North Holland Publ. Co., Amsterdam.

Rackham, O (1976) *Trees and woodland in the British landscape* Dent, London.

Roberts, H.A. (1970) Viable weed seeds in cultivated soils, *Rep. Natn. Veg. Res. Stn*, (1969) 23–38.

Roberts, H.A. (1979) Periodicity of seedling emergence and seed survival in some Umbelliferae, *J. Appl. Ecol.*, **16**, 195–201.

Roberts, H.A. and Feast, P.M. (1973) Emergence and longevity of seeds of annual weeds in cultivated and undisturbed soil, *J. Appl. Ecol.*, **10**, 133–43.

Roberts, H.A. and Ricketts, M.E. (1979) Quantitative relationships between the weed flora after cultivation and the seed population in the soil, *Weeds Res.*, **19**, 269–75.

Robinson, R.G. (1949) Annual weeds, their viable seed population in the soil, and their effect on yields of oats, wheat and flax, *Agron. J.*, **41**, 513–8.

Root, R.B. (1973) Organisation of a plant–arthropod association in simple and diverse habitats: the fauna of collards, *Ecol. Monogr.*, **43**, 95–124.

Roughton, R.D. (1962) A review of literature on dendrochronology and age determination of woody plants, *Colo. Dep. Game Fish Tech. Bull.*, **15**, 99pp.

Roughton, R.D. (1972) Shrub age structures on a mule deer winter range in Colorado, *Ecology*, **53**, 615–25.

Salisbury, E.J. (1942) *The reproductive capacity of plants*. Bell and Sons, London.

Sarukhan, J. (1974) Studies on plant demography: *Ranunculus repens* L., *R. bulbosus* L. and *R. acris* L: II. Reproductive strategies and seed population dynamics, *J. Ecol.*, **62**, 151–77.

Sarukhan, J. (1977) Studies on the demography of tropical trees, Ch. 6, pp. 163–82 in Tomlinson, P.B. and Zimmerman, M.H. (eds), *Tropical trees as living systems*, Cambridge University Press, Cambridge and New York.

Sarukhan, J. (1980) Demographic problems in tropical systems, Ch. 8, pp. 161–88 in Solbrig, O.T. (ed.), *Demography and evolution in plant populations*. Blackwell, Oxford; University of California Press, California.

Sarukhan, J. and Gadgil, M. (1974) Studies on plant demography: *Ranunculus repens* L., *R. bulbosus* L. and *R. acris* L: III. A mathematical model incorporating multiple modes of reproduction, *J. Ecol.*, **62**, 921–36.

Sarukhan, J. and Harper J.L. (1973) Studies on plant demography: *Ranunculus repens* L., *R. bulbosus* L. and *R. acris* L: I. Population flux and survivorship, *J. Ecol.*, **61**, 675–716.

Schaffer, W. (1974) Optimal reproductive effort in fluctuating environments, *Am. Nat.*, **108**, 783–90.

Schaffer, W.M. and Gadgil, M.D. (1975) Selection for optimal life histories in plants, Ch. 6, pp. 142–56 in Cody, M.L. and Diamond, J.M. (eds), *Ecology and evolution of communities*. Belknap Press, Cambridge Mass. and London, England.

Schaffer, W.M. and Schaffer, M.V. (1977) The adaptive significance of variations in reproductive habit in the *Agavaceae*, Ch. 22, pp. 26–276 in Stonehouse, B. and Perrins, C.M. (eds), *Evolutionary ecology*. Macmillan, London.

Schaffer, W.M. and Schaffer, M.V. (1979) The adaptive significance of variations in reproductive habit in the Agavaceae: II.—Pollinator foraging behaviour and selection for increased reproductive expenditure, *Ecology*, **60**, 1051–69.

Scott, P.R. Johnson, R. Wolfe, M.S. Lowe, H.J.B. and Bennett, F.G.A. (1980) Host-specificity in cereal parasites in relation to their control, *Appl. Biol.*, **5**, 349–93.

Sharitz, R.R. and McCormick, J.F. (1975) Population dynamics of two competing annual plant species, *Ecology*, **54**, 723–40.

Shaw, M.W. (1968) Factors affecting the natural regeneration of sessile oak (*Quercus petraea*) in North Wales. I. A preliminary study of acorn production, viability and losses, *J. Ecol.*, **56**, 565–83.

Sheldon, J.C. and Burrows, F.M. (1973) The dispersal effectiveness of the achenepappus units of selected compositae in steady winds with convection, *New Phytol.*, **72**, 665–75.

Silvertown, J.W. (1980a) The evolutionary ecology of mast seeding in trees, *Biol. J. Linn. Soc.*, **14**, 235–50.

Silvertown, J.W. (1980b) Leaf-canopy induced seed dormancy in a grassland flora, *New Phytol.*, **85**, 109–18.

Silvertown, J.W. (1981a) Microspatial heterogeneity and seedling demography in species rich grassland, *New Phytol.*, **88**, 117–28.

Silvertown, J.W. (1981b) Seed size lifespan and germination date as co-adapted features of plant life history, *Am Nat.*, **118**, p. 860–64.

Smith, A.P. and Palmer, J.O. (1976) Vegetative reproduction and close packing in a successional plant species, *Nature*, **261**, 232–3.

Smith, F.G. and Thorneberry, G.O. (1951) The tetrazolium test and seed viability, *Proc. Ass. Off. Seed Analysts N. Am.*, **41**, 105–8.

Snell, T.W. and Burch, D.G. (1975) The effects of density on resource partitioning in *Chamaesyce hirta* (Euphorbiaceae), *Ecology*, **56**, 742–6.

Snoad, B. (1981) Plant form, growth rate and relative growth rate compared in conventional, semi-leafless and leafless peas, *Scient. Hort.*, **14**, 9–18.

Soane, I.D. and Watkinson, A.R. (1979) Clonal variation in populations of *Ranunculus repens*, *New Phytol.*, **82**, 537–73.

Sohn, J.J. and Polikansky, D. (1977) The costs of reproduction in the mayapple *Podophyllum peltatum* (Berberidaceae), *Ecology*, **58**, 1366–74.

Solbrig, O.T. and Simpson, B.B. (1974) Components of regulation of a population of dandelions in Michigan, *J. Ecol.*, **62**, 473–86.

Solbrig, O.T. and Simpson, B.B. (1977) A garden experiment on competition between biotypes of the common dandelion (*Taraxacum officinale*), *J. Ecol.*, **65**, 427–30.

Stearns, F.W. (1949) Ninety years change in a northern hardwood forest in Wisconsin, *Ecology*, **30**, 350–8.

Stearns, S.C. (1976) Life history tactics: a review of the ideas, *Quart. Rev. Biol.*, **51**, 3–47.

Stearns, S.C. (1977) The evolution of life history traits, *Ann. Rev. Ecol. Syst.*, **8**, 145–71.

Stephenson, A.G. (1979) An evolutionary examination of the floral display of *Catalpa speciosa* (Bignoniaceae), *Evolution*, **33**, 1200–9.

Summerhayes, V.S. (1968) *Wild orchids of Britain.* 2nd edn, Collins, London.

Swain, A.M. (1973) A history of fire and vegetation in northeastern Minnesota as recorded in lake sediments, *Quat. Res.*, **3**, 383–96.

Symonides, E. (1977) Mortality of seedlings in natural psammophyte populations, *Ekol. Pol.*, **25**, 635–51.

Symonides, E. (1978) Numbers, distribution and specific composition of diaspores in the soils of the plant association spergulo-corynephoretum, *Ekol. Pol.*, **26**, 111–22.

Symonides, E. (1979a) The structure and population dynamics of psammophytes of inland dunes: I. Populations of initial stages, *Ekol. Pol.*, **27**, 3–37.

Symonides, E. (1979b) The structure and population dynamics of psammophytes on inland dunes: II. Loose-sod populations, *Ekol. Pol.*, **27**, 191–234.

Symonides, E. (1979c) The structure and population dynamics of psammophytes on inland dunes: III. Populations of compact psammophyte communities, *Ekol. Pol.*, **27**, 235–57.

Tamm, C.O. (1956) Further observations on the survival and flowering of some perennial herbs: 7. *Oikos*, **7**, 274–92.

Tamm, C.O. (1972a) Survival and flowering of some perennial herbs: II. The behaviour of some orchids on permanent plots, *Oikos*, **23**, 23–8.

Tamm, C.O. (1972b) Survival and flowering of perennial herbs: III. The behaviour of *Primula veris* on permanent plots, *Oikos*, **23**, 159–66.

Tansley, A.G. (1917) On competition between *Galium saxatile* L. (*G. hercynium* Weig.) and *Galium sylvestre* Poll. (*G. asperum* Schreb.) on different types of soil, *J. Ecol.*, **5**, 173–9.

Thomas, A.G. and Dale, H.M. (1974) The role of seed reproduction in the dynamics of established populations of *Hieracium floribundum* and a comparison with that of vegetative reproduction, *Can. J. Bot.*, **53**, 3022–31.

Thompson, K. and Grime, J.P. (1979) Seasonal variation in seed banks of herbaceous species in ten contrasting habitats, *J. Ecol.*, **67**, 893–921.

Thompson, P.A. (1975) Characterization of the germination responses of *Silene dioica* (L.) Clairv. populations from Europe, *Ann. Bot.*, **39**, 1-19.

Trenbath, B.R. (1974) Biomass productivity of mixtures, *Adv. Agron.*, **26**, 177–210.

Trenbath, B.R. and Harper, J.L. (1973) Neighbour effects in the genus *Avena*: I. Comparison of crop species, *J. Appl. Ecol.*, **10**, 379–400.

Tripathi, R.S. and Harper, J.L. (1973) The comparative biology of *Agropyron repens* L. (Beav.) and *A. caninum* L. (Beav.): 1. The growth of mixed populations established from tillers and from seeds, *J. Ecol.*, **61**, 353–68.

Tubbs, C.R (1968) *The new forest: an ecological history*. David and Charles, Newton Abbot, Devon.

Turkington, R.A. Cavers, P.B. and Aarssen, L.W. (1977) Neighbour relationships in grass-legume communities: I. Interspecific contacts in four grassland communities near London, Ontario, *Can. J. Bot.*, **55**, 2701–11.

Turkington, R.A. and Harper, J.L. (1979) The growth, distribution and neighbour relationships of *Trifolium repens* in a permanent pasture: 4. Fine-scale biotic differentiation, *Can. J. Bot.*, **57**, 245–54.

Van Valen, L. (1975) Life, death, and energy of a tree, *Biotropica*, **7**, 260–9.

Varley, G.C. Gradwell, G.R. and Hassell, M.P. (1973) *Insect population ecology*. Blackwell, Oxford; University of California Press, USA.

Vasek, F.C. (1980) Creosote bush: long-lived clones in the Mojave desert, *Am. J. Bot.*, **67**, 246–55.

Waloff, N. and Richards, O.W. (1977) The effect of insect fauna on growth, mortality and natality of broom *Sarothamnus scoparius*, *J. Appl. Ecol.*, **14**, 787–98.

Waters, W.E. (1969) The life table approach to an analysis of insect impact, *J. For.*, **67**, 300–4.

Watkinson, A.R. (1978a) The demography of a sand dune annual *Vulpia fasiculata*: II. The dynamics of seed populations, *J. Ecol.*, **66**, 35–44

Watkinson, A.R. (1978b) The demography of a sand dune annual *Vulpia fasiculata*: III. The dispersal of seeds, *J. Ecol.*, **66**, 483–98.

Watkinson, A.R. and Harper, J.L. (1978) The demography of a sand dune annual *Vulpia fasiculata*: I. The natural regulation of populations, *J. Ecol.*, **66**, 15–33.

Watson, M.A. (1979) Age structure and mortality within a group of closely related mosses, *Ecology*, **60**, 988–97.

Watt, A.S. and Fraser, G.K. (1933) Tree roots and the field layer, *J. Ecol.*, **21**, 404–14.

Weaver, S.E. and Cavers, P.B. (1979) The effects of emergence date and emergence order on seedling survival rates in *Rumex crispus* and *R. obtusifolius*, *Can. J. Bot.*, **57**, 730–8.

Welbank, P.J. (1963) A comparison of competitive effects of some common weed species, *Ann. Appl. Biol.*, **51**, 107–25.

Werner, P.A. (1975) Predictions of fate from rosette size in teasel (*Dipsacus fullonum* L.), *Oecologia*, **20**, 197–201.

Werner, P.A. (1979) Competition and coexistence of similar species, Ch. 12, pp. 287–310 in Solbrig, O.T., Jain, S. Johnson, G.B. and Raven, P.H. (eds), *Topics in plant population biology*. Macmillan, London and New York.

Werner, P.A. and Caswell, H. (1977) Populations growth rates and age vs. stage-distribution models for teasel (*Dipsacus sylvestris* Huds.), *Ecology*, **58**, 1103–111.

Werner, P.A. and Platt, W.J. (1976) Ecological relationships of co-occurring golden rods (*Solidago*: compositae), *Am. Nat.*, **110**, 959–71.

Wesson, G. and Wareing, P.F. (1969a) The role of light in the germination of naturally occurring populations of buried weed seeds, *J. Exp. Bot.*, **20**, 402–13.

Wesson, G. and Wareing, P.F. (1969b) The induction of light sensitivity in weed seeds by burial, *J. Exp. Bot.*, **20**, 414–25.

West, N.E. Rea, K.H. and Harniss, R.O. (1979) Plant demographic studies in sagebrush-grass communities of southeastern Idaho, *Ecology*, **60**, 376–88.

Whelan, R.J and Main, A.R. (1979) Insect grazing and post-fire plant succession in south-west Australian woodland, *Aust. J. Ecol.*, **4**, 387–98.

Whipple, S.A. (1978) The relationship of buried, germinating seeds to vegetation in an old growth colorado sub-alpine forest, *Can. J. Bot.*, **56**, 1505–9.

White, J. (1979) The plant as a metapopulation, *Ann. Rev. Ecol. Syst.*, **10**, 109–45.

White, J. (1980) Demographic factors in populations of plants, Ch. 2, pp. 21–48 in Solbrig, O.T. (ed.), *Demography and evolution in plant populations*. Blackwell, Oxford; University of California Press, California.

White, J. (1981) The allometric interpretation of the self-thinning rule. *J. Theor. Biol.*, **89**, 475–500.

White, J. and Harper, J.L. (1970) Correlated changes in plant size and number in plant populations, *J. Ecol.*, **58**, 467–85.

White, P.S. (1979) Pattern, process, and natural disturbance in vegetation, *Bot. Rev.*, **45**, 229–99.

Whitehead, F.H. (1971) Comparative autecology as a guide to plant distribution, pp. 167–76 in Duffey, E.O. and Watt, A.S. (eds), *The scientific management of animal and plant communities for conservation*. 11th Symp. Brit. Ecol. Soc., Blackwell, Oxford; F.A. Davies and Co., Philadelphia, USA.

Whitmore, T.C. (1977) Gaps in the forest canopy, Ch. 27, pp. 639–55 in Tomlinson, P.B. and Zimmerman, M.H. (eds), *Tropical trees as living systems*. Cambridge University Press, Cambridge and New York.

Wilbur, H.M. (1976) Life history evolution of seven milkweeds of the genus *Asclepias*, *J. Ecol.*, **64**, 223–40.

Wilbur, H.M. (1977) Propagule size number, and dispersion pattern in *Ambystoma* and *Asclepias*, *Am. Nat.*, **111**, 43–68.

Willey, R.W. and Heath, S.B. (1969) The quantitative relationships between plant population and crop yield, *Adv. Agron.*, **21**, 281–321.

Williams, O.B. (1970) Population dynamics of two perennial grasses in Australian semi-arid grassland, *J. Ecol.*, **58**, 869–75.

Williamson, M.H. (1972) *The analysis of biological populations*. Arnold, London.

Willson, M.F. and Bertin, R.I. (1979) Flower-visitors, nectar production and inflorescence size of *Asclepias syriaca*, *Can. J. Bot.*, **57**, 1380–8.

Willson, M.F. and Price, P.W. (1977) The evolution of inflorescence size in *Asclepias* (Asclepiadaceae), *Evolution*, **31**, 495–511.

Willson, M.F. and Rathcke, B.J. (1974) Adaptive design of the floral display in *Asclepias syriaca* L., *Am. Mid. Nat.*, **92**, 47–57.

Wit, C.T. de (1960) On competition, *Versl. Landbouwk. Onderz.*, **66**, 1–82.

Wright, H.E. Jr and Heinselman, M.L. (eds) (1973) The ecological role of fire in natural conifer forests of western and northern America, *Quat. Res.*, **3**, 317–513.

Wyatt, R. (1976) Pollination and fruit-set in *Asclepias*: a reappraisal, *Am. J. Bot.*, **63**, 845–51.

Wyatt, R. (1980) The reproductive biology of *Asclepias tuberosa*: I. Flower number, arrangement, and fruit-set, *New Phytol.*, **85**, 119–31.

Wyatt Smith, J. (1958) Seedling/sapling survival of *Shorea leprosula*, *S. parviflora* and *Koompassia malaccensis*, *Malay For.*, **21**, 185–93.

Yakovlev, M.S. and Zhukova, G.Y. (1980) Chlorophyll in embryos of angiosperm seeds, *Bot. Notiser*, **133**, 323–36.

Yarranton, G.A. (1966) A plotless method of sampling vegetation, *J. Ecol.*, **54**, 229–37.

Yoda, K. Kira, T. Ogawa, H. and Hozumi, K. (1963) Self-thinning in overcrowded pure stands under cultivated and natural conditions, *J. Biol. Osaka City Univ.*, **14**, 107–29.

Index

Note: Plant and animal species are entered under their Latin names followed by the common name in parentheses, e.g. *Populus nigra* (black poplar), where both names are given in the text.